God, Science, and Designer Genes

God, Science, and Designer Genes

An Exploration of Emerging Genetic Technologies

Spencer S. Stober and Donna Yarri

PRAEGER
An Imprint of ABC-CLIO, LLC

A B C 🝔 C L I O

Santa Barbara, California • Denver, Colorado • Oxford, England

Library of Congress Cataloging-in-Publication Data

Stober, Spencer S.
   God, science, and designer genes : an exploration of emerging genetic technologies / Spencer S. Stober and Donna Yarri.
      p. cm.
   Includes bibliographical references and index.
   ISBN 978-0-313-35253-9 (hard copy : alk. paper) — ISBN 978-0-313-35254-6 (ebook)
   1. Genetic engineering. 2. Human gene mapping. 3. Human Genome Project. I. Yarri, Donna. II. Title.
   TP248.6.S74   2009
   660.6'5—dc22   2009011113

13   12   11   10   09   1   2   3   4   5

This book is also available on the World Wide Web as an eBook.
Visit www.abc-clio.com for details.

ABC-CLIO, LLC
130 Cremona Drive, P.O. Box 1911
Santa Barbara, California 93116-1911

This book is printed on acid-free paper ∞

Manufactured in the United States of America

To Lois Stober and David Westcott

"Science without religion is lame.
Religion without science is blind."
—Albert Einstein

CONTENTS

PREFACE

Human genetics has certainly come a long way since Gregor Mendel, appropriately called the father of genetics, worked with pea plants in the 19th century. Many of us remember from our high school biology classes the diagrams that demonstrated the mechanisms of heredity not only in pea plants but also in all species, representing a major step forward in our understanding of how heredity operates in living organisms. This major leap forward in science has significant implications for the lives of living creatures. The significance of and vibrant discussion surrounding emerging genetic technologies, however, have certainly been heightened by the completion of the Human Genome Project (HGP).

The initial goal of the HGP was to map the entire human genome. The race ended in 2003, when both the private and the public sectors published their results nearly simultaneously. We now know not only the approximate number of genes in the human body but also, more important, numerous specific genes associated with known diseases. Mapping the human genome, certainly an impressive scientific achievement, is leading to technology developments that have the potential to affect human generations for the foreseeable future and which raise troubling moral questions for which we have no easy answers.

Anticipating the difficult issues that would arise from the science and technology surrounding the human genome, the National Institutes of Health (NIH) allotted approximately 3 to 5 percent of the HGP budget specifically for serious consideration of the ethical, legal, and social implications (the ELSI Project) of the HGP. The intention of the ELSI Project was to ensure that serious discussion about developments in genetics be undertaken. Important topics include the ones we consider in this book: designing our children, cloning, stem cell research, the relationship between genes and behavior, genetic privacy, and the social justice implications of all of these.

Our involvement in working in genetics began when the directors of the Ethics Institute at Dartmouth College, Ronald M. Green and Aine Donavan, applied for and received a grant on the ELSI Project from the NIH. With the funding they conducted a series of workshops over a 3-year period with scholars and professionals interested in some of these moral issues in genetics. We were two of the scholars who applied for and received the grant within the grant to attend one of their workshops, which we did in June 2004, at Howard University in Washington, D.C. The purpose of Dartmouth's grant was to bring together a diverse group who would first, learn about HGP and ELSI; second, engage in discussion about the issues raised; and third, make plans to disseminate the information in the communities from which we came. Working within different disciplines (Spence in biology and Donna in theology) enabled us not only to understand and approach the subject from the perspective of our unique disciplines but also to frame and understand the issues in a truly interdisciplinary way. It is our hope that this volume will enable you to do the same.

ACKNOWLEDGMENTS

A work such as this goes well beyond the world of simply the authors of the book. Along the way there have been numerous individuals who have offered their assistance and support, and without whom this manuscript never would have been conceived, let alone completed.

We are especially grateful to Ronald M. Green and Aine Donavan at Dartmouth College. Their foresight in realizing the significance of the completion of the Human Genome Project and the recognition of the importance of having educators and scholars explore the ethical, legal, and social implications of this project led us to participate in their workshop funded by a grant obtained through the National Institutes of Health. They have continued to support our work in this area, and we remain conversation partners on this very important topic.

We were also encouraged in our work by numerous people at our institution, Alvernia University. We want to especially thank Dr. Gerald Vigna, our former dean, who has been one of our strongest advocates. He wrote a very supportive letter to help us obtain our grant, and has continued to cheer us on throughout the subsequent years of our research together. Dr. Kevin Godfrey and Ms. Elaine Schalck, our former department chairpersons, also encouraged us in our work. Ms. Bobbie Rohrbach worked tirelessly to meet our multiple requests for interlibrary loan materials, and Jimmy Haluska, a work study student, obtained numerous articles on religion and genetics. Finally, our President Thomas Flynn and our Provost Shirley Williams are to be commended for their recognition of the importance of scholarly activity and in their efforts to actively support our efforts.

We also thank our original editor, Suzanne Staszak-Silva, without whose initial proactive interest in our work this book would not have been written. In addition, we appreciate all of the hard work put in by the numerous people at Praeger/Greenwood and ABC-CLIO involved in all the stages of the completion of our manuscript.

Finally, thank you to all of our family, friends, and colleagues who believed in us and who supported us in various ways.

PART I

Laying the Foundations: Biology, Theology, and Being Human

1

In the Beginning: Introductory Issues

Learn from me, if not by my precepts, at least by my example, how dangerous is the acquirement of knowledge, and how much happier that man is
who believes his native town to be the world, than he who aspires to
become greater than his nature will allow.
— Dr. Victor Frankenstein in Mary Shelley, *Frankenstein*

Mary Shelley's well-known classic, *Frankenstein*, created the myth of
the Frankenstein monster. Fair or not, this tale has come to stand as
a metaphor for the hubris of science, and of the scientist, which
both oversteps its boundaries and whose experiments can have unintended, dire, and even fatal consequences. The single-minded pur-
suit of the scientist in search of knowledge prohibits him from
considering other possible implications.

Dr. Victor Frankenstein is a scientist fascinated with the concept
of the reanimation of life. With the help of his assistant, he gathers
body parts from various dead individuals, fashions a creature of gi-
gantic proportions, and then animates him by harnessing the energy
of lightning. The obsession with which he works to make this
creature stands in stark contrast to the revulsion he feels when
he actually comes face-to-face with this "monster." Although
Dr. Frankenstein laments what he has done, the monster is forced
to fend for himself in a hostile world, which is simultaneously
repulsed and frightened by his demeanor. The despair and sadness
the monster feels because of the rejection and ostracism he experiences at the hands of society turn to murderous rage, focused particularly on the cherished friends of Dr. Frankenstein, who is the true
object of the monster's anger. The tale concludes with the death of
both Dr. Frankenstein and his creation, whose fates are inextricably
linked forever.

Although we may not yet be able to create life from dead matter, recent developments in the field of genetics, especially genetic engineering, raise the specter again of Frankenstein's monster. We are on the precipice of exciting new scientific discoveries and technological possibilities. Might we be entering a world of no return, in which our desire to improve the human condition may actually backfire? Will science be our salvation, or our undoing?[1] But these are not questions simply for the scientists or for those who apply the science in technological innovations; they are for all of us. In fact, we maintain that science is not the "enemy." Science is simply one of many disciplines that has its own unique tasks and whose inquiry and results are not necessarily problematic in themselves. Rather, it is what we do with the insights and knowledge that we gain which will set our course as a species and as a world.

Yet the Frankenstein myth serves as a cautious reminder that science unencumbered by ethical considerations can be problematic at best and devastating at worst. The new emerging genetic technologies might force us to conclude that "not everything which is technologically possible can be considered morally permissible."[2] However, we think that this discussion needs to happen in an interdisciplinary way, drawing on the wisdom and insight in particular of the disciplines of science and religion and, for the purposes of this volume, more specifically from the perspectives of biology and Christian theology.

What has precipitated the renewed interest in genetics and its implications has been the Human Genome Project (HGP). In this chapter, we first offer some observations about genetic information, describe the HGP and describe the race for its completion, summarize its goals, indicate some its benefits and subsequent importance, and raise some implications and areas of concern. We then include some of the same information for the coevolved Ethical, Legal, and Social Implications (ELSI) Project. Finally, we provide an overview of the subsequent book chapters.

All life forms, dead or alive, contain genetic information. Genetic information abounds. We can gather it, then make copies, and put it into living libraries. We can identify its function and then examine and modify the information. We can move the information from one life form to the next and then clone these contemporary organisms. Genetic science has provided the techniques for the development of these technologies. To put it another way, scientists can sequence complete genomes and build gene libraries; they can identify and modify genes and their function; they can cut and splice DNA into vectors for insertion into organisms;[3] and they can clone genes, cells, and organisms. All these techniques were developed since Watson and Crick elucidated the structure of DNA in 1953.[4] Now, more than 50 years later, we have achieved a major scientific breakthrough—the HGP. In some ways, the developments of 1953 pale in comparison to what we have now achieved and will eventually achieve in the future. In fact, the HGP has been referred to as the third "great" scientific project of the 20th (and now 21st) century. The first was the Manhattan Project in the 1940s, which produced a test bomb and ultimately two bombs that were dropped on Japan (cost: $22 billion). Clearly, the proliferation of nuclear warheads is not a "great" achievement, but nuclear energy potentially could sustain our future. The second was the Apollo Project, from approximately 1963 through the 1970s, whose goal was to place humans on the moon (cost: $95 billion). The third is the Human Genome Project, spanning the years from 1990 to 2003.[5] Congress originally estimated a total cost of $3 billion for the 13-year HGP project, but, believe it or not, it took less time and money than expected. The project was finished in 2 years at an estimated cost of $2.7 billion.[6] So what exactly is the HGP?

The HGP refers to the 13-year effort coordinated by the U.S. Department of Energy (DOE) and the National Institutes of Health (NIH), along with several international partners. The project results are widely available through numerous Web sites, articles, and books.[7] This project had three discrete stages, each with its own plan: from 1990 to 1995 (revised in 1993 because of unexpectedly rapid progress), from 1993 to 1998, and from 1998 to 2003, with the plans for the latter two becoming increasingly detailed. Although the HGP is officially finished, analyzing the data will go on, as will developing technologies drawing on these data and sequencing the genomes of other creatures.

The HGP is as much a goal accomplished as it is a new beginning. But the path to its accomplishment was one that can best be characterized as a race between the private and public sectors. In 1990, when the U.S. government assigned the task of mapping the human genome to the DOE and the NIH, they formed a partnership called the International Human Genome Sequencing Consortium, with the goal to map the complete genome in 15 years. James

D. Watson served as the first director and was replaced by Francis Collins 2 years later. David B. Resnik, in *The Price of Truth*, illustrates the tension as private and public sectors raced to map the genome.[8] We briefly summarize his observations: The science and some of the technologies were in place, but several key events hastened the process. The consortium's approach used well-developed clone by clone sequencing techniques, but J. Craig Venter (while working for the NIH and the consortium) proposed a faster shotgun sequencing approach that was ultimately rejected by the consortium. Venter then left the public consortium to establish in 1992 a not-for-profit private organization, The Institute of Genomic Research (TIGR). Six years later, Venter and Applera Corporation joined to form a private, for-profit company called Celera Genomics and relied on Venter's shotgun sequencing. Resnik identifies the 3-year period between 1998 and 2001 as a time when the public and private interests were competing. Public sector scientists were depositing their newfound sequences in public databases, but Celera scientists did not share their data, hoping to recover their research and development investment by selling access to the data. The race was on. Celera ultimately revised their business plan, and both groups published draft versions of the human genome in 2001 (the consortium in *Nature* and Celera in *Science*). The HGP race was finished 2 years ahead of schedule when final versions from both groups were published in 2003. Goal accomplished!

The HGP was complete, but it was apparent from the start that this was a project with significant potential to affect society and science. The effect on society was addressed primarily through the ELSI Project. What was the impact on science? During Darwin's time, biology could be categorized as more of an inductive science, a theory arising from overwhelming observations. Darwin's theory then guided our inquiry as we took a more deductive approach and developed hypotheses for testing and refining the theory. The rise of modern genetics, the HGP, and the resulting proliferation of genomic data for humans and other species constitute another new beginning for biologists. We are once again trying to make sense of overwhelming amounts of information. Although the focus has been on the human genome, the genomes of more than 180 organisms have been sequenced since 1995 (including yeast, fruit flies, mice, and dogs, to name a few).[9] Believe it or not, we have more than 3,600 genes in common with plants.[10]

It is important for our purposes to understand what a genome is. The term "genome" is used to represent the total amount of DNA contained in a cell or organism.[11] DNA is a molecule that can store and transmit information in a chemical format (coded in combinations of the following molecules: A, adenine; C, cytosine; G, guanine; and T, thymine). The DNA in cells and organisms contains these molecules in specific sequences (genes) that are replicated and passed from cell to cell, and generation to generation, with a fairly high degree of (but not perfect) accuracy. Each gene has a specific location (locus or address) in the total sequence of DNA for a particular genome. Collectively within a species, the genes carry all the necessary information, including mechanisms for adaptation, to sustain a species. The human genome was originally sequenced using a collection of DNA strands from several individuals, and although this was a complete sequence (including known as well as yet to be identified genes), it does not include all the variability within our species.

A significant outcome of genomic science today is that it is easier to map genomes than it is to understand what all the information means. Scientists are not sure why, but most DNA is made up of noncoding sequences ("junk") scattered within genes and among genes in the genome of a species. To appreciate the difficulty scientists face, imagine that a genome is the "book of life" for a species. Scientists are attempting to decode a language (DNA) that has only four letters (A, C, G, and T) in a book of directions (genes) that contains 23 chapters (pairs of human chromosomes). The ordered list of letters (the sequence of the genome) is complete, but now the challenge is to make sense of the information. Scientists are working to identify the meaningful directions (genes and their functions) among nonsense words (noncoding sequences) in each chapter (chromosomes). What surprises are in store for scientists and society? It remains to be seen how this new information will affect current evolutionary theory and our understanding of life in general as well as what it means to act in an ethical manner with regard to these new developments. It is the HGP that gave rise to these new developments.

The best way to define the HGP is by examining its goals—what it actually hoped to achieve. The overall general purpose of the HGP was to better understand genetic factors in human diseases, with the ultimate goal of developing new strategies for diagnosis, treatment, and prevention. The HGP had six basic goals. The first was to identify all of the approximately 20,000 to 25,000 genes in human DNA. This is a considerably smaller number than the original estimate of 100,000 genes, and continues to be refined. Up-to-date information on human genes as they are discovered can be retrieved at the Online Mendelian Inheritance in Man (OMIM) Web site.[12] The second goal was to determine the sequences of the 3 billion chemical base pairs that make up human DNA. The sequence of chemical base pairs is important because that is how DNA stores information for transmission and expression. The third goal was to store this information (the chemical base sequences) in databases. Genomic information can be stored digitally and within genomic libraries. Public access was and continues to facilitate rapid scientific achievements in this field. GenBank, sponsored by the National Center for Biotechnology Information (NCBI), is a public database of digital sequences.[13] Genomic libraries store functional DNA in living organisms. The fourth goal was to improve the tools for analysis, which continues. Bioinformatics evolved as a discipline to support and manage the vast amount of genomic data being generated, and so did the field of computational biology, as scientists began to formulate questions about what the data means. These areas of biology are ripe with future opportunities for new biological insights. The fifth goal was to transfer related technologies to the private sector. From the very beginning, this project, which was coordinated by U.S. government agencies, was intentionally designed to incorporate the work of the private research laboratories as well. The sixth goal was to address the ethical, legal, and social issues that would certainly arise from this project. The year 2003 represented the achievement of these goals, but the work needs to continue.

The actual and potential benefits of the HGP are far-reaching, affecting many areas of science and having implications for human health, environmental issues, criminal justice (e.g., DNA forensics with regard to identification), and agriculture, to name just a few. It already has created and will continue to create a deeper understanding of human biology and its relationships to other species, in terms of both similarities and differences. It is the impact on human health, however, that primarily concerns the average citizen. Within the area of human health, there are at least four important benefits. The first is the discovery of genes related to about 2,000 diseases. Probably 3,000 more still need to be discovered. The second is the possibility of individuals having their own genome sequenced; this now is prohibitively expensive, but the NIH eventually hopes to make it available at a cost of not more than $1,000. The third is the timely detection of genes causing problems, and the greater availability of genetic tests for embryos/fetuses, newborns, and adults, particularly those with a family history of certain diseases. Finally, targeting drugs and treatments to individuals based on their unique genome will allow for the possibility of personalized medicine.

The specific questions related to the individual technologies are addressed in the chapters on these topics. But some broader questions have been raised by the HGP. Should there be limits on science or the application of science? What does it mean to be human? Are we "playing God"? In other words, are we as humans overstepping our bounds? Who will be the winners and the losers with regard to new genetic technologies? What implications are there for society with regard to nonmedical corrections such as intelligence, athletic ability, and personality traits? How can we best engage in interdisciplinary dialogue, and who will be invited to the table? And for all of the foregoing and subsequent questions, who will be making the decisions? The kinds of questions that have been raised regarding the HGP demonstrate that the originators of this project were very wise in creating the ELSI Project alongside the HGP to address these issues. It is to the ELSI Project that we now turn our attention.

We are beginning to realize that the effect on society of these new scientific and technological developments may be much greater than previous discoveries in genetics. For that reason, when the HGP was started, the DOE and the NIH set aside 3 percent and 5 percent of their respective allocated funds to map the human genome. This represented the world's largest bioethics program. These funds were earmarked for projects that consider the ethical, legal, and social implications of human genome research. ELSI coevolved with the HGP; most recently, some ELSI considerations include the study of human genetic variation, the effect of genomics on health care, the awareness of gene-environment interactions, and how racial, ethnic, and socioeconomic factors affect policy and genetic services. ELSI aims "to explore how new genetic knowledge may interact with a variety of philosophical, theological, and ethical perspectives."[14] It was designed to create an interdisciplinary dialogue that would engage professionals from the relevant areas to consider the effect on society and individuals as these technologies were increasingly developed and became available to the wider population.

The goals for the ELSI Project were part of the HGP goals from the very beginning. In the first stage (1990–1993), ELSI was simply one of seven areas to be addressed. In the plan for the second stage (1993–1998), ELSI goals were more clearly enunciated:

- Continue to identify and define issues and develop policy options to address them.
- Develop and disseminate policy options regarding genetic testing services with potential widespread use.
- Foster greater acceptance of human genetic variation.
- Enhance and expand public and professional education that is sensitive to sociocultural and psychological issues.[15]

The goals enunciated in the 1998–2003 plan were even more detailed:

- Examine issues surrounding the completion of the human DNA sequence and the study of human genetic variation.
- Examine issues raised by the integration of genetic technologies and information into health care and public health activities.
- Examine issues raised by the integration of knowledge about genomics and gene-environment interactions in non-clinical settings.
- Explore how new genetic knowledge may interact with a variety of philosophical, theological, and ethical perspectives.
- Explore how racial, ethnic, and economic factors affect the use, understanding, and interpretation of genetic information; the use of genetic services; and the development of policy.[16]

The ELSI Project is very important for a number of reasons. First, it is significant that the ethical concerns coevolved with and were a part of a scientific enterprise from its inception, probably a first. Second, the coordinators of HGP anticipated that the implications and impact for society would be deep and far-reaching, and they were correct in their assessment. Third, it allowed for an interdisciplinary and intentional dialogue to consider the issues that would arise. Fourth, since ethics is part of theology (as well as of some other fields), although it may not have been an expressed goal, it enabled religious traditions to contribute to the discussion. Finally, it demonstrates that disciplines cannot easily be distinguished and separated, and that scientific discoveries and technological innovations in genetics must have the support of society to ultimately make a difference.

One of the important by-products of the ELSI Project has been to raise questions for discussion. In addition to the general conceptual questions raised earlier, the ELSI Project is concerned with questions in several specific areas.[17] One area is fairness in the use of genetic information. "Who should have access to personal genetic information, and how will it be used?" Who should make decisions as to who should have access? Another area is privacy and confidentiality of genetic information. "Who owns and controls genetic information?" What mechanisms can or should be put in place to guard genetic information? A third area is "psychological impact and stigmatization due to an individual's genetic differences. How does personal genetic information affect an individual and society's perceptions of that individual? How does genomic information affect members of minority communities?" How do we distinguish between normal human variation and disabilities? A fourth area is "reproductive issues, including adequate informed consent for complex and potentially controversial procedures, use of genetic information in reproductive decision making, and reproductive rights." What are the larger societal issues raised by new reproductive technologies? "How reliable and useful is fetal genetic testing?" A fifth area is clinical issues for health care providers, patients, and the general public. "How do we prepare health care professionals for the new genetics? How do we prepare the public to make informed choices? How do we as a society balance scientific limitations and social risk with long-term benefits?" A sixth area is "uncertainties associated with genetic tests for susceptibilities and complex conditions linked to multiple genes and gene-environment interactions. Should testing be performed when no treatment is available?" Should testing ultimately be done for all individuals? A seventh area is conceptual/philosophical concerns. "Do people's genes make them behave in certain ways? ... Where is the line between medical treatment and enhancement?" A final area is the commercialization of products, including property rights, patents, and accessibility of data and materials. Should life forms be able to be patented? Obviously, no one volume can address all of these issues, but this list of questions shows some of the breadth and depth of the issues involved. In our chapters on specific issues, we attempt to address some of these concern related to those actual and potential technologies.

The importance of the ELSI Project will certainly continue. There are several basic areas for future development of this project. First, funding will continue to be available for ELSI work, including new studies, interdisciplinary analysis, and support for the development of courses, workshops, and conferences. Second, it is likely that the proliferation of ethics centers devoted exclusively to the new genetics will continue, and it is also likely that other more general ethics centers as well as academic organizations will examine these issues. Finally, the steady stream of publications devoted to these issues from theological, philosophical, legal, and public policy perspectives will certainly expand. It is especially heartening that religion has become an active voice engaged in these discussions, as it tends to do with regard to medical ethics in general.

This book is divided into three broad sections. Part I has three chapters; its purpose is to lay out some of the foundational issues that are important when considering issues in genetic science before moving on to the specific issues in Part II. Particular focus is on how the disciplines of biology and theology can contribute to the wider discussion by appreciating the contributions of each in exploring what it means to be human. Chapter 1 is an introduction to the book in general. Chapter 2 explores the individual nature of and the possible relationships between science and religion in general and theology and biology in particular. First, we describe each discipline using comparison and contrast of issues such as their ways of knowing, their realms of exploration, their tasks, and their tools. Second, we examine the various kinds of relationships that have been conceptualized for religion and science (e.g., are they independent disciplines, are they interdependent, is science supreme, is religion supreme). Third, we offer our view of how these disciplines should be related (we argue that both disciplines are essential for assessing genetic issues). Finally, we discuss the implications of this for issues in genetic science.

Chapter 3 introduces concepts in biology and theology necessary for looking at the specific issues in Part II. We define important scientific concepts that underlie genetics and demonstrate their importance for the HGP. We also describe the significance of genetic science, indicating the opportunities and challenges. In addition, we explain basic concepts in ethics and describe tools for analysis of some of these difficult and ever-emerging issues. We consider both secular and Christian theological approaches to ethics.

Part II discusses several of the more controversial topics within genetic science that we think are of interest to many. We begin each chapter with a relevant ethical dilemma (a short hypothetical example, in italics) that introduces some of the tough moral questions raised by this kind of dilemma, briefly describe the technology currently available, and examine some of the ethical questions. We conclude by attempting to balance the perspectives with practical considerations for humans. In the description of each of the chapters that follow, we identify some ethical dilemmas raised by the current and future technology.

Chapter 4 explores the implications of the fact that we now have the technology to manipulate our genes. This is one of the most important and controversial topics, because it deals with the possibility of having some control over the creation of our children. We discuss sex selection, choosing enhancements and choosing against disabilities, and even the possibility of choosing for disabilities (such as a deaf couple choosing to have a deaf child). We describe the difficulties inherent even in defining what a disability is as well as discuss the hidden messages in terms such as "designer baby," "test tube baby," and "genetic engineering."

Chapter 5 examines specifically the therapeutic nature of cloning. Most people think of embryos when hearing the term "stem-cell research," but there are actually two kinds of stem cell research: adult stem-cell research and embryonic stem-cell research. This chapter defines what is meant by stem-cell research in general, describes the scientific process behind it, distinguishes between the two types, and considers the potential benefits and concerns of using this technology. Certainly embryonic stem-cell research is the more controversial, so more of the chapter is focused on this area, including the promise in many arenas of human health, as well as the ethical concerns such as the moral and physical status of embryos, and the subsequent relationship to the abortion issue.

Chapter 6 considers the concept of cloning and why it is a term that many people have heard of but do not fully understand. We define this ambiguous concept, explain where and how it already exists in nature, and consider how cloning might be of value and yet detrimental to humans. In addition, we examine the nature/nurture debate, which is often excluded in discussions of cloning. The issue of cloning actually has two components: reproductive and therapeutic cloning. In this chapter we focus on the former.

Chapter 7 discusses in greater detail the nature/nurture debate—how much of who we are is a matter of our nature (genes) and how much is a matter of our environment (nurture). Most would agree that both are important influences in the makeup of individual humans and that both are helpful in distinguishing humans from each other. However, with the advent of genetic technologies, some researchers are looking for genes for some behaviors that in the past would more often have been attributed to human free will or to strong environmental influences. In this chapter, we summarize the research on genetic predispositions to particular behaviors and explore the implications of such genes actually existing.

Chapter 8 considers growing concerns about how to protect individual privacy in light of developing technologies. This is an especially important issue when it comes to one's DNA. We offer an overview of the technology available with regard to DNA analysis, as well as potentially constructive and harmful uses of one's DNA information. Important issues include whether a database with everyone's DNA profile should exist; determining who would have access to such a database (e.g., medical personnel, the government, one's family, one's insurer, or one's employer); how to avoid discrimination, particularly in the workplace and with regard to insurance coverage; and concerns with genetic profiling.

Part III contains the book's conclusions in Chapter 9. Genetic technologies are here to stay, and are likely to produce increased sophistication and new developments in the future. But a deeper social question is about how we will distribute fairly the benefits and burdens of these new technologies. Hence, we must consider the issue of justice. We assess the overall opportunities represented by emerging genetic technologies; the ethical challenges that exist now and potential challenges for new technologies; and how all of this affects the question of social justice, including who will have access to these new technologies and how might these technologies might affect other areas of social justice.

NOTES

1. For an excellent discussion on the problems of science going beyond its bounds, see Mary Midgley, *Science as Salvation: A Modern Myth and Its Meaning* (London and New York: Routledge, 1992).

2. Pontifical Council for Pastoral Assistance, *Charter for Health Care Workers* (Boston: St. Paul Books & Media, 1999), par. 44.

3. Vectors behave like the legendary Trojan horse, but instead of carrying soldiers into the city of Troy, vectors carry DNA into an organism's cells. For example, viruses can be loaded with relevant genetic information for delivery into cells. The advantage of using viruses is that they infect specific tissues and can be modified (attenuated) so as not to cause disease while delivering the DNA, but this process is not always as precise as is sounds, and unintended consequences do arise.

4. James D. Watson and Francis H. C. Crick, "Molecular Structure of Nucleic Acids: A Structure for Deoxyribose Nucleic Acid," *Nature* 171 (April 25, 1953): 737–38.

5. Hessel Bouma III, "Completing the Human Genome Project: The End Is Just the Beginning," *Perspectives in Science and Christian Faith* 52, no. 3 (2000): 152–55. Because he was writing in 2000, he did not yet know the endpoint for the HGP.

6. http://www.genome.gov/11006943.

7. Some of the most important and helpful websites on the HGP and the ELSI Project are www.ornl.gov, the website of the Oak Ridge National Laboratory, which is managed by the U.S. Department of Energy; www.genome.gov, which is part of the NIH website; www.ncbi.nlm.nih.gov, the website of the National Center for Biotechnology Information, and www.hugo-international.org, the website of the Human Genome Organisation. Most of the information in the sections on the HGP and the ELSI Project come from www.ornl.gov.

8. David B. Resnik, *The Price of Truth: How Money Affects the Norms of Science* (New York: Oxford University Press, 2007), 14–18.

9. A complete list is available through the Genetic News Network at http://www.genomenewsnetwork.org/resources/sequenced_genomes/genome_guide_p1.shtml (accessed August 25, 2008).

10. Ricki Lewis, *Human Genetics: Concepts and Applications*, 8th ed. (Boston: McGraw-Hill, 2008), 429.

11. Throughout this book, important scientific terms are defined when they are first used. An outstanding resource for readers is *Talking Glossary of Genetic Terms* sponsored by the National Human Genome Research Institute, one of the National Institutes of Health (NIH), at http://www.genome.gov/glossary.cfm?search=genome (accessed August 25, 2008).

12. The Online Mendelian Inheritance in Man (OMIM) can found at http://www.ncbi.nlm.nih.gov/sites/entrez.

13. The GenBank website at http://www.ncbi.nlm.nih.gov/Genbank (accessed July 20, 2008).

14. Sandy B. Primrose and Richard M. Twyman, *Genomics: Applications in Human Biology* (Malden, MA: Blackwell Publishing, 2004), 22; for additional information see http://www.genome.gov/10001618 (accessed March 29, 2008).

15. www.ornl.gov/sci/techresources/Human_Genome/project/5yrplan/5yrplanrev.shtml (accessed July 22, 2008).

16. www.ornl.gov/sci/techresources/Human_Genome/project/hg5yp/goal.shmtl (accessed July 22, 2008).

17. The following section incorporates some complete sentences from www.ornl.gov/sci/technresources/Human_Genome/elsi/elsi.shtml (accessed July 8, 2008), and is supplemented by our own observations.

2

Playing God, or Using the Brains God Gave Us? Biology and Theology in Dialogue

"All the same," said the Scarecrow, "I shall ask for brains instead of a heart;
for a fool would not know what to do with a heart if he had one."
"I shall take the heart," returned the Tin Woodman; "for brains do not make
one happy, and happiness is the best thing in the world."
Dorothy did not say anything, for she was puzzled to know which of her two
friends was right.. . .

L. Frank Baum, *The Wonderful Wizard of Oz*

INTRODUCTION

In L. Frank Baum's book, *The Wonderful Wizard of Oz*, which most
of us know from watching the magnificent film adaptation, the Tin
Woodman and the Scarecrow discuss which is more important—the
head or the heart. Although this dichotomy is certainly not unique
or original to this book, and we want to avoid an overly simplified
and problematic schema in which science refers to the head and religion to the heart, this distinction highlights the importance of different ways of knowing. It can also highlight the significance of
approaching complex issues through an interdisciplinary lens.

When considering issues in genetics, it is very important to
understand the broad areas of religion and science as well as the specific disciplines of theology and biology. We can read about and discuss the emerging genetic technologies with regard to only one
discipline, but it is crucial to understand how both areas contribute
to a broader understanding of the topic, including the challenges
and possibilities. Thus, it is possible to talk about how cloning
works, the knowledge of which lies within the domain of science. It
is also possible to ask what the implications are for our concept of

God if humans can clone, especially human cloning, the belief of which lies within the domain of religion. But we all will be better able to examine the ethical and theological concerns if we understand the science behind the technology. We cannot, however, stop at the science either. Most of us are primarily concerned with the implications raised by the science and subsequent technological developments of the new genetics.

In this chapter, we describe each of the disciplines introduced here, and then compare and contrast theology and biology, looking at their different realms of exploration, assumptions, ways of knowing, tasks, and tensions. In some cases we make the connection to genetic science. Next we consider models, or paradigms, that lay out the possible ways in which the relationship between religion and science can be conceptualized. We discuss how we think the relationship between these two disciplines should be conceived (ultimately, that both disciplines are essential for assessing genetic issues) and offer some concluding comments.

A DESCRIPTION OF EACH DISCIPLINE

Two important sets of distinctions need to be defined: religion and science, and theology and biology. Religion and science are the broader categories, and theology and biology are specific fields within those categories; they are usually the concrete areas in which scholars and professionals work. Therefore, in most of the book, we emphasize theology and biology.

Religion is a nearly universal human experience, present in all cultures and at all times. It has been defined in numerous ways, and not all ways have been favorable. For example, Sigmund Freud thought religion to be comparable to a childhood neurosis, and Karl Marx considered it to be "the opiate of the people." Recent critics of religion include Richard Dawkins, Christopher Hitchens, and Kai Nielsen. But most of the definitions have been positive, and have been honest attempts to get at the essence of the divine nature and its relationship to humans primarily and even the rest of creation. Friedrich Schliermacher defined religion as a feeling of absolute dependence. Rudolf Otto defined it as that which grows out of and gives expression to the experience of the holy. Alfred North Whitehead defined religion as what an individual does with his or her solitariness.

Although religion may be difficult to define, it ultimately focuses on the meaning and purpose of life. Three general questions are raised by virtually all religions: Where do we come from? How shall we live? What will ultimately happen to us? Of course, most religious traditions believe in some kind of God, or gods, although that is not necessarily true (e.g., in certain forms of Buddhism). But religion is a study of ultimate reality, which usually includes God, humans, and the entire created world. It is a uniquely human venture (as far as we know) and apparently a universal and abiding dimension of human experience. This search for ultimate meaning manifests itself in numerous religious traditions: Eastern religions, Western religions, and indigenous traditions, to name a few broad categories. To understand a religion, though, one must ultimately explore its theology.

Theology is related to the concept of religion. It is actually the study of the beliefs, practices, rituals, history, and so on of a specific religious tradition. Thus, every religion, even every smaller religion within a religion (e.g., Methodism within Christianity), has its own unique theology. Theology is reflection on religious faith; a famous definition of Christian theology by an early church father, Anselm, which is often quoted, is that theology is "faith seeking understanding." Theology is a bridge, in a sense, between belief and cognition. Theology is the attempt to express a religious faith in language.

But it is very important to understand that the term theology always needs a modifier. Thus, although we can define theology in a general way, theology is ultimately about trying to understand a particular religious tradition. Therefore, we cannot ask what theology teaches, but we can ask what Christian theology teaches, what Hindu theology teaches, what Muslim theology teaches. In addition, each major world religion also has many subgroups, and what distinguishes one group from another, and in fact what usually caused the split from a previous group, was a difference of theology on one or numerous points.

The focus in this book is on Christian theology. The Christian tradition can broadly be divided into Catholic, Orthodox, and Protestant (and the last subdivided into numerous varieties, called denominations). All groups within Christianity agree on a core group of theology, or beliefs, but they obviously differ as well. In the appropriate chapters, we discuss what Christian theology as a whole, for example, believes about God and human beings, and at other times we distinguish between, for example, what the Catholic Church believes about embryonic stem cell research and positions that some other traditions may hold. When we use the term theology, however, it should be understood to mean Christian theology in general unless otherwise specified.[1]

Science is a process of inquiry that enables us to better understand the natural world. The famous biologist Edward O. Wilson wrote in *Consilience: The Unity of Knowledge* that "science is neither a philosophy nor a belief system."[2] Scientists consider the natural world to be the realm of science. A primary assumption is that humans are capable of understanding the natural world because it is consistently organized according to laws governing energy and matter, and that these aspects of the natural world are subject to observation and experimentation. The methods of science have proven to be very effective in describing the natural world. Scientists gather empirical evidence to construct theories and models that organize their knowledge of the natural world and to further guide their process of inquiry.

Scientists develops methods of inquiry that, when applied appropriately, produce valid and reliable information. The conventional academic branches of science such as chemistry, physics, and biology take these methods very seriously—so much so that the very content that they generate must be verifiable and, if not, the process of inquiry is revised. If the outcomes are verified, but conflict with prevailing theory, then the theory is modified to fit observations (e.g., evolutionary theory). This is not to say that bias does not exist, but humans have found these methods and the information generated to be very reliable in its many applications, and as a result, human expectations for science have escalated to the point that science is viewed by many as the only way to understand the natural world. Scientists, with their keen minds and well-developed methods, make every effort to accurately describe the natural world, but scientists are only human, and in their hearts they may, if so inclined, accept things on faith (that is, without empirical evidence). Francis S. Collins, director of the HGP, came to his faith in God while practicing medicine. When he was asked to direct the HGP, he first said no, but after serious contemplation he said yes, because, in his words, "Here was a chance to read the language of God, to determine the intimate details of how humans had come to be."[3]

It is also important to distinguish between science and technology. For some, the HGP is what is sometimes called basic science, with its primary goal to enhance our understanding of the natural world. For others, the HGP is what is sometimes called an applied science, with research to support specific technical applications. Both basic and applied sciences can provide a theoretical basis for the development of techniques with practical applications. We adapt these techniques to develop technologies that may, or may not, improve the human condition. A scientific understanding of the natural world, and then knowing how to intervene in what is natural, sets the stage for the development of technological applications. Genetic science is giving us many emerging genetic technologies. Whether or not we are "playing God" is a question on the minds of many people. This raises questions regarding scientific intent and human actions. Should science, a human endeavor, seek this knowledge? Are we capable of causing no harm with this knowledge? That said, the "HGP genie" is already "out of the bottle," and biologists now have a responsibility, as with global warming, to participate in societal discourse regarding these technologies. If we are reading "the language of God," as Francis Collins believes, then biology, along with future genomic discoveries, might be the scientific discipline that forces us to take a leap of faith.

Biology is the study of life. Biologists, as do scientists in general, hold worldviews informed by their understanding of theory and processes for their respective subdisciplines: cell biology, genetics, microbiology, and ecology, to name a few. Darwin's evolutionary theory was a turning point for the science of biology. The theory itself evolved from a process of contemplation by Darwin and others of vast amounts of information they gathered primarily by observations of life in all of its complexity. This inductive scientific approach resulted in evolutionary theory, which now frames nearly all questions for consideration in biology today. Thus it can be said that biology became a deductive science along with Darwin's theory. Modern-day biologists with formal training in the physical sciences (or so-called hard sciences) now seek to explain life using the experimental methods of chemistry and physics. This approach has proven to be very effective in molecular biology, with significant outcomes in modern medicine and genetics. But now we are paying a great deal of attention to genes, assuming that they can explain the origins of life and ongoing adaptations. We are also able to manipulate genes with a view toward improving our futures, though this is a subject of considerable discussion.

One of the best ways to further biological understanding, as well as theological understandings, is to explore its worldviews. A worldview is simply one's way of looking at the world—it includes things such as implicit or hidden assumptions, explicit assumptions, the kinds of questions one brings to any theoretical or even practical discussion, how one gathers knowledge, and even what one believes is knowable. The worldview of a discipline is expressed in its methodology, which includes the components just mentioned. Methodology refers to the ways that disciplines go about doing their research. We recognize that theology and biology, as two distinct and unique disciplines, each have their own methodology with which they approach questions and issues. To better understand how they each operate and how they can inform each other's disciplines, especially with regard to genetic science, we explain how each discipline approaches five broad areas that contribute to its methodology and to its worldview.

The realm of inquiry of theology is the physical and the metaphysical, the natural and the supernatural, this world and the next world. Although the physical refers to the actual literal created order, the world in which we and all other inhabitants of the planet live, there is a particular emphasis within theology on human beings. Thus, theology inquires about how we got here, the character of the one who brought us here (namely, God), the essence of human nature, and how humans should morally live with regard to God and to the rest of creation. The last is particularly the focus of moral theology, and is our emphasis in subsequent chapters with regard to genetic technologies. However, theology also has metaphysical inquiries, which includes the nature of God, the concept of the spiritual in general, and certainly questions about the afterlife. A foundational belief of traditional theology is that there is another dimension of the world, which we can experience even now within the physical, such as spiritual or mystical experiences in relationship with God (through prayer or rituals, for example) or for some believers, through encounters with other spiritual beings such as angels and saints. Theology assumes, though, that there is a connection between the physical and the metaphysical, specifically with regard to morality. Thus, how one lives one's earthly life can affect one's future destiny. Perhaps the afterlife can be conceptualized, as C. S. Lewis does in his book,

The Great Divorce, as an extension of the way humans conduct them-selves on their earthly sojourn. Thus, for those who eventually go to heaven, for example, their earthly experience would have been a foretaste of this. The point is that for theology, the empirical physi-cal world that we see is not all that there is. To bring the physical and the metaphysical together, an important way of describing theol-ogy's realm of inquiry is to say that it deals with the "is" and the "not yet."

The realm of inquiry for biology is the natural world, and whereas theology may seek to include a metaphysical realm, that is not the case for biology. Biologists assume that life is governed by the univer-sal laws of physics and chemistry, and that we can understand every-thing about all life processes in terms of these laws. But not all biological phenomena are easily explained in terms of these universal laws, such as the role of nurture in the expression of genes. Although chemical techniques enable us to map genomes and describe gene function, we are now surprised to find that we share many more genes than expected with other life forms. Chemical and physical laws are not adequate to explain these observations. Modern bio-chemical techniques can describe the structure and function of genes. But questions such as the origins of life and how we relate to other life forms are not easily answered. It remains to be seen how far the realm of inquiry for biology will expand.

Theology starts out with basic assumptions. The first and foremost is that God exists. Although arguments for the existence of God, which have usually been developed by theologians, often are dis-cussed in introductory classes on theology, they are not considered necessary to convince the faithful. Theology does not believe that the existence of God needs to be proven, even if it could be done. In many ways, faith is seen as more of a "leap of faith," as main-tained by Sören Kierkegaard. However, this does not mean that believers need to "check their brains at the door," since even taking the leap means that you are putting your faith in some particular God and in some particular understanding of this God. Theology assumes certain things about the nature and character of God as well. Thus, God is considered the ultimate ground of being, the uncreated Creator who has brought all things into existence (this is not necessarily contradictory to the notion of evolution), who is morally good, and who has created humans in God's image. What it means to be created in the image of God has been understood both ontologically (with regard to human nature) or functionally (with regard to the role humans play with respect to the rest of creation). Ontologically, it can mean that humans are created with free will; functionally it can mean that humans have dominion/stewardship over the rest of creation. Theology also assumes that this God, as a moral being, desires that humans live moral lives, but that they are ultimately free to make their own choices. Theology assumes that God reveals to humans information on how they should live morally.

Evolutionary theory is the primary assumption (or prevailing paradigm) that guides inquiry in the science of biology, and a second assumption is that life is constrained by physical laws governing the material world. In contrast to theology, the existence of God is not an assumption for the discipline of biology. Evolutionary theory is used to organize knowledge gathered by scientific methods in an effort to explain the origins and ongoing adaptation by species to their many environments on earth. A recent outcome of evolutionary theory, enhanced by modern genetics, is a wave of genetic reductionism. In its extreme form, the reductionist view holds that genes are the fundamental unit of life, and that life's origins, adaptations, and interactions with the environment can therefore be explained entirely in genetic terms. This view is expanding to include behaviors, as some social scientists along with biologists seek to explain human behavior in genetic terms (a form of "genetic determinism").

Steven Rose, an experimental biologist with a gift for taking the larger view, calls for a perspective that transcends genetic reductionism. He calls for a view of life that is organism-centered, not gene-centered.[4] This position does not discount the value of evolutionary theory, nor does it imply that life is independent of physical laws that govern the material world. But a commonly stated truism applies here: "Give a kid a hammer and everything is a nail." To make sense of this, consider the fact that most biologists are well-trained in chemistry and its experimental approaches, and thus it should be no surprise that these techniques become the tools of first choice for many biologists. Chemistry and molecular biology have flourished since Darwin's time, and like the child with a hammer, biologists soon considered the biochemical aspects of the gene to be fundamental in their explanations of nature. Genes are fundamental to our understanding of life, but they are not the whole picture.

Steven Rose offers several ways that biology as a discipline can be made "whole again." He recommends that biologists in their study of nature should return to an emphasis on observation and contemplation, with less emphasis on intervention; that is, to examine the whole organism in its context. He asks biologists to consider how the history of their discipline has shaped human knowledge of the natural world, and to acknowledge that their level of analysis is dependent on their disciplinary view. Rose suggests that active intervention in living systems in order to understand the primary parts, and then attempting to explain the whole, ignores the fact that as biological systems emerge and evolve, the systems themselves influence the development and action of the parts.[5] This is a very important consideration when examining the interaction of genes and the environment.

Theology has its own unique ways of knowing. One way to understand how a discipline creates knowledge and approaches reality is to explore what it uses as its sources or basis for knowledge. The sources of knowledge for theology are primarily four: Scripture, tradition, reason, and experience. It is important to note that what results in differences in theology and specifically morality among Christian denominations has to do with which sources they emphasize, their order of importance, and any additional sources they might include.[6] The four sources just mentioned were developed within the Methodist or Wesleyan tradition (called the "quadrilateral" and developed by Outler, an early Methodist theologian). It is important to define what is meant by these sources.

Most religious traditions have scriptures, or sacred writings; for Christianity, these sacred writings exist as the Bible. There are certainly other religious writings that can help us grow in our understanding of and living out of one's faith, but the Bible is foundational. Christian theology maintains that knowledge about God as well as guidelines on how to live can be found in the Bible. Tradition refers to the teaching of the church that is passed on through successive generations, and includes development of the interpretation of the Bible in light of church teaching over time. Thus, it is insufficient simply to say that the "Bible says thus and such"; we must also see how the church interpreted and continues to interpret what the Bible "says." Reason refers to the idea that God has imbued humans with the intelligence to make moral decisions and to think cognitively. In fact, throughout Christian history, reason has been described as the characteristic that distinguishes humans from other animals. Reason can also include insights and knowledge from other disciplines, including science.[7] Finally, experience refers to human experiences and subsequent reflection on these experiences that might help us to understand God and human life in relationship to God better. In essence, though, theology believes strongly in the concept of revelation: that God has chosen to be known and self-discloses God's self to humans in nature, human history, and even within the conscience of each individual. These sources provide us not only with information about the nature of God and humans but also with insights on how we should live morally, which are the focus of the next chapter.

Biology shares its primary way of knowing with the other scientific disciplines: the reasoned application of scientific methods that are designed to yield valid and reliable information to model our view of the natural world. Biology, unlike theology, does not consider revelation to be a reliable way of understanding our natural world. Scientific methods are a source of strength for biology as well as a limitation. The validity of biological insight is improved by methods that rely on testability. This is easier for branches of biology such as molecular genetics, where variables can be measured and subjected to controlled experiments to demonstrate causal relationships among the parts (such as how a defective gene may result in a defective hormone receptor on a cell surface). In contrast, this is not so easy for branches of biology such as behavioral genetics, where the variables are not easily measured (such as a predisposition to violence). An association (or correlation) between two variables is not absolute proof that they are in a direct causal relationship. When direct causal relationships cannot be demonstrated between variables such as genes and behavior, then biologists (and social scientists) use statistical tests as an indicator of how strongly these variables are associated with one another. All too often we say we "know" something in biological terms as if a causal link has been established, when instead only a very high correlation has been demonstrated. Is this "knowing"? The answer depends on a discipline's standard for proof. Biological knowing relies on both causation and correlation, and its methods should consider the opportunities and limitations of each approach. It remains to be seen if biology, along with the other sciences, is able to complete our understanding of the natural world.

Theology has its own tasks, or questions that it tries to address. All religions are concerned with the origins, morality, and future of human life. The major task of theology is to understand God and subsequently how humans should live in relationship to this God. In fact, theology *is* the "study of God." But that definition is too broad, or at least needs to be explicated because understanding the character of and having a relationship with this personal God has implications for humans in terms of how they should live. Within the broad task of theology, then, we can talk about subsequent tasks. One task of theology is to determine what God is like. This is done primarily with reference to the sources previously described. Another task is to reflect on the implications of the nature of God for human action in the world, with regard to God, other humans, and the rest of the created order. With regard to God, theology can offer guidelines for rituals and other religious practices that might enable one to better relate to and have experiences of this God. With regard to humans, it can help us live morally in such a way that we treat others as also created in the image of God, with the dignity that comes along with that. With regard to the rest of creation, it can provide guidelines for understanding how the different aspects of the created world relate, and can enable us to live in such a way as to be good stewards of God's creation.

Biology's primary task is to enhance our understanding of life forms and their environments. In an ideal world, this task is carried out as research to increase our understanding, but not necessarily with practical applications as a goal. In the real world, this is usually not the case, and applications then become a second task. Recall that Francis Collins was asked to direct the HGP and agreed because he realized that this was an important opportunity to enhance our understanding of life. In addition, he realized the practical implications. Collins soon found himself consumed by a race between the private and public sectors. He became concerned that the race would diminish the significance of the human genome, so he ultimately struck a deal for simultaneous announcements of the first draft even though, as Collins reports, the private sector's genome sequence was derived from more than 50 percent public data. For Collins, this became a battle of ideals; he asked, "Would the human genome sequence, our shared inheritance, become a commercial commodity, or a universal public good?"[8] Thus, biologists cannot overlook the application of knowledge they seek. In fact, some would argue that it is a primary task for biology to improve the human condition, and others would argue that this is beyond the realm of biology. In either case, biological sciences are a human endeavor and, as such, choices are made to publicly fund or not to fund expensive scientific endeavors such as mapping genomes. Public funding calls for consensus on the perceived appropriateness of such research, including some oversight (as we see with the ELSI Project and the HGP). If public funding is unavailable, then incentives exist for scientists to seek private funding, and in some cases, without the public good in mind.[9] It remains to be seen how this will impact the development of emerging genetic technologies.

Finally, theology has tensions. A major area of tension for theology is how God can operate within a causally closed universe. It basically gets at the question of how God is involved with or in the world, if at all. Deism assumes that there was a Creator at the beginning of the world, but that this Creator simply set things in motion and "permits" things to happen. Theism, however, the category to which Christian theology belongs, believes in a personal God who not only set things in motion but also continues to uphold, sustain, and influence the world. But how does this God operate? Is the world determined or not? To what extent are humans free to change things? Where does the role of God end and the role of humans begin? Different theological traditions may differ on how God operates in the world, and this certainly has implications for genetic science. It especially arises with regard to genetic technologies, where many are asking, "Are we playing God?"

Because biology is a discipline that relies heavily on the scientific approach to inform its understanding of life, one might conclude that the major tensions are methodological. But they are not; biologists make every effort to work through the methodological issues. Instead, for this biologist, a tension within the discipline of biology is whether or not biology should seek divine causes in addition to natural causes to explain life, and what the implications of this might be. Is there a public perception that scientists are too objective and thus lack "heart," much like the Tin Woodman in *The Wonderful Wizard of Oz*? For many scientists, a public perception that we lack a heart is disconcerting; most scientists view themselves as humans first and scientists second. E. O. Wilson said that "scientists are to science what masons are to cathedrals."[10] This analogy can also serve our purposes here. The mason can cut and lay stone to build a cathedral, and with each stone the cathedral takes the expected form. The scientist can process and contribute scientific knowledge, and with each bit of information, a theory may or may not accurately serve as a more precise model for nature. Both the mason and the scientist may from time to time set their task aside to reflect on their work, and while the mason should have a fairly good idea of what the cathedral is going to look like, this is not necessarily the case for a scientist studying natural phenomena; hence an inevitable tension exists.

It is obvious that the disciplines of theology and biology approach reality quite differently and even have unique ideas as to what constitutes reality. However, it is also important to emphasize that they are both recognized as significant disciplines in their own right, with their own unique mode of inquiry and worldview. But the broader question that has been a significant one throughout human history, and which continues to receive considerable attention, is what the proper relationship between these two disciplines should be. Although the discussion is usually framed with regard to the broader areas of religion and science, what is said certainly applies to biology and theology, and subsequently has important implications for discussions about the new genetics.

MODELS/PARADIGMS FOR COMPARING RELIGION AND SCIENCE

Because each discipline approaches reality from different perspectives and uses different methodologies, it is important to consider how these disciplines might relate to each other. There is a perception in the minds of many modern people that religion and science have always had a conflictual relationship. This perception has been fueled by famous examples of individuals such as Galileo and Darwin, whose advances in science appear to have created a stir in the faithful. On the face of it, scientific advances can often threaten long-held and cherished theological ideas. For example, Galileo was imprisoned for defending Copernicus' idea that the sun and not the earth was the center of the universe. For some religious people, this challenged the notion of humans as the center of the universe. The findings of Darwin about evolution challenged human superiority in a different way, given the idea that humans may have evolved not only similarly to other animals but might even have descended from them and hence shared common ancestors. It also challenged the biblical account of the origins of life, in which God created the world in six days and created all the individual species. Although many people do believe that religion must defend itself against in particular the theory of evolution (consider the ongoing debate in many school districts as to whether or not "creationism" should be taught in the science classrooms in public schools), many other religions and individuals have come to see evolution as complementary to their faith, as long as God was involved in the process. This is to say that the too simple model of religion and science as being enemies is much more complex than it may seem to be on the surface. Throughout history most of the advances in science have been met simultaneously with rejection and acceptance by religious people, and the scientific community has simultaneously consisted of both individuals with deeply held religious convictions and those who have been agnostics or atheists.[11]

However, despite this complexity, it is still helpful to examine models that attempt to conceptualize different possible relationships between religion and science. One of the most commonly used ones is that offered by Ian G. Barbour, who offered a fourfold typology, in which he maintains that all four models have existed throughout history. The first model is that of conflict, in which religion and science view each other as enemies. Science sees religion as the enemy to new scientific knowledge because of adherence to particular religious beliefs, and religion sees science as challenging these traditional beliefs. Biblical literalists who maintain that the theory of evolution conflicts with religious faith, and scientists who claim that the scientific evidence for evolution is incompatible with any form of theism, are both examples of the conflict model.

The second model is independence, in which the distinct nature of the two disciplines is acknowledged and maintained. There are two possible versions to this model: the first views them as dealing with different aspects of reality, whose unique languages and concerns (science with how things work and objective facts, and religion with values and ultimate meaning) keep them in separate camps, and thus conflict is not even possible. The second version sees them as offering complementary perspectives on the world that are not mutually exclusive.

The third model is dialogue, in which the two disciplines can respect each other's area of expertise, can enter into discussion especially with regard to boundary questions such as why the universe is orderly and intelligent, or can compare methodologies, acknowledging both the differences and similarities.

The fourth model is integration, in which there is a more systematic partnership between the disciplines in order to understand reality and discoveries from a more holistic perspective. Examples could be astronomers acknowledging that physical constants in the early universe could be there by design, or some within religious traditions maintaining that some of its beliefs should be reformulated in light of science.[12]

Barbour has been criticized by some for offering an overly simplified model of the relationship between religion and science. Gregory Cantor and Chris Kenny, while acknowledging Ian Barbour as probably the most cited author in the area of religion and science, also take issue with his four-fold typology. They think it is problematic that he tries to fit all cases into his schema, and that it may not be applicable particularly to earlier periods in history. They also maintain that his models incorporate values and commitments that the unwary reader may not be aware of. They think that his rejection particularly of the conflict model does not take seriously the writings of people such as Richard Dawkins and other advocates of the conflict thesis, and does not take seriously the conflicts that have been an almost continual backdrop to this relationship. They also challenge the notion that science and religion are necessarily "bounded" activities, and question whether they can be restricted in this way. Finally, they maintain that his models, while perhaps applicable to Christianity, may not be applicable to other religious traditions.[13]

Other individuals studying science and religion have developed their own typologies, a few of which are briefly noted. Some have slightly adapted Barbour's model, but others make more substantive changes. John Polkinghorne provided two additional categories to Barbour's model, accepting the first four; his fifth and sixth categories are consonance (in which both disciplines retain their autonomy while attempting reconciliation in areas of overlap) and assimilation (achieving the maximum possible conceptual merging of the disciplines).[14] John Haught created his own four-fold schema, the first two of which are similar to Barbour's—Haught's conflict corresponds to Barbour's conflict, whereas Haught uses the term contrast instead of Barbour's independence. Haught's contact is similar to Barbour's dialogue, and Haught's confirmation differs somewhat from Barbour's integration in that Haught emphasizes the ways in which religion can be deeply supportive and nourishing of the scientific enterprise, particularly with regard to the inherent rationality of the universe.[15] Ted Peters established an eight-fold model: scientism (only science provides the knowledge we know), scientific imperialism (science seeks to cover territory originally possessed by theology), ecclesiastical authoritarianism (theology given authority because of divine revelation), scientific creationism (creationists wanting to make their case in the arena of science), two-language theory (keeping the disciplines in their separate domains), hypothetical consonance (seeking those areas where there is a correspondence between disciplines), ethical overlap (the need of theologians to respond to challenges posed by our technological society), and new age spirituality (emphasizing a holism that overcomes dualisms).[16] Thus, as all of these attempts at models demonstrate, the discussion of the relationship between religion and science is controversial, complicated, and quite contemporary.

Given the interdisciplinary nature of our volume and the different disciplines within which we both work, we think it only fair to be straightforward about how we think the disciplines of science and religion, and particularly theology and biology, should relate. Although we acknowledge the excellent work done by others to modify and refine Barbour's model, we also think that he has a very useful tool for conceptualizing these disciplines, particularly in light of genetic technologies. First, we recognize that we all tend to view reality only or primarily through our own lens. If postmodernism has taught us anything, it is that there is no such thing as the objective, neutral observer. The well-known story of the blind men and the elephant illustrates this point very well. Each man was touching a different part of the elephant—one his tail, one his trunk, one his leg—and yet they all tried to describe the reality of the elephant in light of their limited perception. As individuals trained in particular disciplines, we know that we bring certain assumptions, ways of knowing, and different questions that affect our way of understanding reality. Second, we consider it very important that in order to address the ethical issues raised by the new genetic technologies, we must understand the actual science, the future potential science, and the resulting technologies. Science, as we have seen, does not necessarily set out with practical applications as a goal, and in fact, science for the sake of understanding nature is often far ahead of potential applications. Sometimes society is unaware of potential applications because of an apathy or lack of understanding of the science in its early stages. For example, before we can engage in public discourse on the implications of gene therapy, it is important that we understand the science behind this type of therapy.

Third, with regard to Barbour's typology, we reject the conflict and independent models. The conflict model sets up too sharp a dichotomy and does not recognize the important insights that we can learn from each other. The independent model sets the disciplines too far apart and does not recognize that there are overlapping questions, realms of inquiry, and even methodological similarities. We fully support the dialogue model and believe that the best way forward with regard to genetic technologies is for the disciplines to truly talk to each other. This dialogue should include a genuine respect for the other, based on the idea that multiple visions of reality will result in a truer picture, and will help us address the difficult issues that lie before us as a species. Finally, we believe that we should move toward the integration model, but we also think it is impossible to fully integrate these disciplines in our lifetime, if ever. Although it might be ideal to envision a world in which knowledge is of one essence, we also think it could ultimately lead to a conflation of the disciplines that would not do justice to either one.

We also are particularly drawn to E. O. Wilson's call for consilience. In *Consilience: the Unity of Knowledge*, Wilson adopts Whewell's use of the term consilience as a way to build a case for unification of the natural sciences and humanities by finding common ground across disciplines.[17] Wilson acknowledges that

> the belief in the possibility of consilience beyond science and across the great branches of learning is not yet science. It is a metaphysical world view, and a minority one at that, shared by only few scientists and philosophers ... The strongest appeal of consilience is in the prospect of intellectual adventure and, given modest success, the value of understanding the human condition with a higher degree of certainty.[18]

Another book by Wilson, *The Creation: An Appeal to Save Life on Earth*, was the subject of a recent Associated Press article by Jeff Barnard. The significance of Bernard's article for this discussion is the fact that it appeared in the religion section of a small city paper with the title "Biologist Extends Olive Branch to Christians."[19] Titles that sell papers reflect what people are interested in knowing about—clearly religion and science are on the minds of many people. Wilson's book takes the form of a passionate letter to a Southern Baptist pastor. This may reflect the fact that he was born and raised as a Southern Baptist, and while no longer of that faith, he claims that the Christian lyrics and spirituality he experienced along with his exploration of nature as a boy were formative experiences that influence him to this day.[20] However, he now describes himself as a secular humanist.[21] Wilson's contributions to our understanding of evolutionary biology while a professor at Harvard and as a science writer today are monumental. Wilson emphasizes that there is a great need for humans to be good stewards of life on earth, and that this is a common ground on which theology and biology must join forces.[22] Wilson is not alone in this request. Francis Collins, HGP director and a devoted Christian, reminds us of two contrasting views of science and religion. On the one hand, materialists triumph in the findings of science that have greatly enhanced our understanding of nature, and there are those that now pronounce God to be a superstition. On the other hand, believers are convinced that spiritual introspection trumps science, and they perceive science to be a threat to God. Collins considers both positions to be dangerous. It is time for a truce. Religion and science should unite to seek "all" truths.[23]

What does all of the foregoing have to do with new genetic technologies? Everything! The new advances in biology, and particularly genetic science, raise questions with which we will all have to grapple. Science will move forward, the technological innovations will follow, and the questions of what we should do as well as the answers we provide will affect all of us. That is why we considered it important to understand how each discipline approaches issues so that we can understand better how they can work in concert. The HGP was completed by science, but the ELSI Project moves the findings and resultant technologies into the ethical realm. To move closer to a consideration of specific issues such as cloning, stem cell research, genetic privacy, and a genetic basis for behavior, we need to examine two areas: to put it simply, the ABCs of genetic science and of moral reasoning. It is difficult to address the implications of HGP research if we do not understand the basic terms and concepts of genetic science. Likewise, we can better engage in ethical reasoning if we understand traditional approaches to ethical thinking. It is to these two areas that we now turn our attention.

NOTES

1. One of the best books on Christian theology has been written by a well-respected Christian historian: Alister E. McGrath, *Christian Theology: An Introduction*, 4th ed. (Oxford and Malden, MA: Blackwell Publishers, 2006).

2. Edward O. Wilson, *Consilience: The Unity of Knowledge* (New York: Alfred A. Knopf, 1998), 45.

3. Francis S. Collins, *The Language of God: A Scientist Presents Evidence for Belief* (New York: Free Press, 2006), 118–19.

4. Steven Rose, *Lifelines: Biology beyond Determinism* (New York: Oxford University Press, 1998), ix–xi.

5. Ibid., 302–9.

6. For example, the Catholic Church uses the natural law tradition.

7. For a fuller discussion of reason in light of religious faith, see Ronald M. Green, *Religious Reason: The Rational and Moral Basis of Religious Belief* (New York: Oxford University Press, 1978). He subsequently wrote an additional work that examines how reason functions in comparative religious traditions: *Religion and Moral Reason: A New Method for Comparative Study* (New York and Oxford: Oxford University Press, 1988).

8. Collins, *The Language of God*, 121.

9. David B. Resnik, *The Price of Truth: How Money Affects the Norms of Science* (New York: Oxford University Press, 2007), provides an excellent and thorough review of this issue.

10. Edward O. Wilson, *The Creation: An Appeal to Save Life on Earth* (New York: W. W. Norton, 2006), 103.

11. For an excellent history of the relationship between science and religion, see John Hedley Brooke, *Science and Religion: Some Historical Perspectives* (Cambridge and New York: Cambridge University Press, 1991).

12. Ian G. Barbour, *When Science Meets Religion: Enemies, Strangers, or Partners?* (New York: HarperCollins, 2000). See also Ian G. Barbour, *Religion in an Age of Science*, Gifford Lectures, vol. 1 (San Francisco: HarperCollins, 1990).

13. Geoffrey Cantor and Chris Kenny, "Barbour's Fourfold Way: Problems with His Taxonomy of Science-Religion Relationships," *Zygon: Journal of Religion and Science* 36, no. 4 (December 2001): 765–81. Ian Barbour responded to these critiques in a subsequent article, "On Typologies for Relating Science and Religion," *Zygon: Journal of Religion and Science* 37, no. 2 (June 2002): 345–59.

14. John Polkinghorne, *Science and Theology: An Introduction* (London and Minneapolis: Fortress Press, 1998).

15. John F. Haught, *Science and Religion: From Conflict to Conversation* (Mahwah, NJ: Paulist Press, 1995).

16. Ted Peters, "Science and Theology: Toward Consonance," in *Science and Theology: The New Consonance*, ed. Ted Peters, 11–39 (Boulder, CO and Oxford: Westview Press, 1998).

17. E. O. Wilson, *Consilience*, 8. Wilson refers to the words of William Whewell in his 1840 synthesis, *The Philosophy of the Inductive Sciences*.

18. Ibid., 9.

19. Jeff Barnard, "Biologist Extends Olive Branch to Christians," *Reading Eagle* (Reading, PA), November 18, 2006, A9.

20. Wilson, *The Creation*, 174–75.

21. Ibid., 3.

22. Ibid., 165. Referring to Wilson's book, Professor Calvin DeWitt at the University of Wisconsin, founder of the Evangelical Environmental Network, stated: "Wilson's book would restore the term creation to scientific discussion." Barnard, "Biologist Extends Olive Branch," A9.

23. Collins, *The Language of God*, 210–23.

3

The ABCs of Genetics and Morality: Basic Concepts in Science and Moral Reasoning

There are mysteries which men can only guess at, which age by age they may solve only in part.

Dr. Van Helsing in Bram Stoker, *Dracula*

INTRODUCTION

As Dr. Van Helsing so aptly observed, there are mysteries of life. In a religious sense, the concept of mystery often is applied to inexplicable things that happen in life and to the character of God which we as finite humans cannot fully understand. In a scientific sense, the concept of mystery can refer to those mechanisms of life that we do not yet or may not ever fully comprehend. In essence, mystery indicates a fundamental barrier that we may not be able to break through; it expresses not only the unknown, but also perhaps even the unknowable. As the Apostle Paul said, "For now we all see through a glass darkly."[1] But continue to guess we all do! The theologian and the scientist, as well as the theologian and scientist within each of us, seek to gain knowledge that will help us to navigate our ever-changing world. With regard to the emerging genetic technologies that result from the science, as well as the ethical questions raised by these technologies, we see that we are still "guessing," and that it is an ongoing process to understand more deeply the "mysteries" of life. Certainly the understanding of genetics has greatly increased over time, and even concepts such as evolutionary theory scientists would not even have been able to guess at 200 years ago, let alone genetically modifying organisms. Similarly with regard to theology and moral reasoning: although theories and ways of thinking about right and wrong have been around virtually since humans have existed, there are new questions that are raised and new answers that need to be discerned that we may only be able to answer in part, and which at one time we had not even considered would be raised. While the answer to whether or not we should premeditatedly kill an innocent person is fairly straightforward from a moral perspective, the question of whether or not we should introduce germ-line modifications into human genes which would irreversibly alter us as a species is a much more complicated issue.

Human life, and in fact all of life, is somewhat of a mystery. And yet it is still important, in order to address the difficult issues that new genetic technologies pose and which are swirling around us, that we attempt to understand the basic concepts in genetic science. This will enable us to see how far we have come in terms of our understanding of genetic mechanisms within all species, how much we do not yet understand, and especially how these basic concepts will enable us to discuss more intelligently and understand more fully the issues that will arise. Thus, we must understand what a gene and a chromosome are before we can explain how they are important to the issue of cloning, for example. In addition, we must understand some traditional approaches to moral reasoning in human history, including theological reasoning, before we can explain how we might best approach the difficult moral choices raised for us by new genetic technologies.

Therefore, we first give an overview of basic concepts in genetics that is a foundation for the discussion to come in subsequent chapters, and we then give an overview of moral reasoning that can be a foundation as we focus on the specific issues in subsequent chapters. We then offer some concluding comments before moving onto the truly controversial notion of "designing our children."

GENES: PAST, PRESENT, AND FUTURE

Our genes tell us a great deal about the origins of life, the form and function of all organisms, and their relationship to each other and their environment. This is powerful knowledge, and it remains to be seen how we will act with this understanding. This section begins with a discussion of evolutionary theory and how genes contain information that is essential for survival of a species. The concept of a gene is considered throughout the discussion, with particular emphasis given to how the genetic information is stored and transmitted from cell to cell and from generation to generation. Examples of gene expression are then considered along with a brief discussion of genetic diversity and its importance.

Evolutionary theory provides a framework for organizing the evidence scientists gather about life and, to a great extent, since Darwin's time, evolutionary theory has guided the direction and process of inquiry in biology. To understand the theory, it is necessary to consider briefly Darwin's perspective. Animal breeding was common practice in Darwin's time, and his observations of animal breeders as they selected for specific traits may have influenced his suggestion that there are natural constraints in nature that could do the same (natural selection as opposed to artificial selection).[2] Darwin's collection of Galapagos finches is the classic example. He noticed a natural variation in beak structure of finches that were isolated on the Galapagos Islands, and he then suggested that this variation served as raw material for natural selection. As food became limited for growing finch populations, breeding success increased for the birds with beaks that could access alternative food sources. Thus, there was natural selection for specialized beaks to access specific foods. Each new species arose over time in a population of finches that could exploit a different food. Isolated first by geography and diet, successful finch varieties were better able to reproduce, and differences accumulated until they became reproductively isolated. That is, they could no longer breed with their living ancestors and produce viable offspring (this is an example of a process that biologists call speciation). The ability to produce viable offspring is a standard used by biologists to distinguish two organisms as belonging to two different species. For example, a donkey and a horse belong to two distinct species that can mate to produce a sterile mule.[3] Since Darwin's time the concept of a gene has come to serve as a fundamental unit to explain the mechanisms for speciation.

Genes endow a species with directions for survival in a particular niche along with other organisms and their environment. These directions are being fine-tuned by natural selection over time and recorded in our genetic language. Genes, like our languages, have evolved over time. Languages store and transmit a culture's learned information for survival; so do genes.[4] Survival in a Darwinian sense is for the "fittest" members of a species, those that are able to attract mates and contribute their successful attributes to the next generation (today, this contribution to the next generation is explained in terms of genes).[5] For example, one would consider the male peacock's huge tail to be an impediment to survival, but when it comes to mate selection, female peacocks are impressed! Over time, strong healthy male peacocks that could garner more resources grew larger tails, and these resources made them more fit for mating. Females that selected mates with larger tails would therefore have the benefit of greater reproductive success than those that opted for the less well-endowed males. Thus, a mate selection dynamic was in place to favor development of larger and larger tail feathers in male peacocks. Generation after generation, genes that facilitated successful mating behaviors, along with larger and larger tail feathers, accumulated in the pool of genes available to members of the peacock species (thus, an attraction for large tail feathers became a genetically programmed instinct in female peacocks). We could provide countless varied examples to support Darwin's theory, but, as with all theories, we must seek explanatory mechanisms. The gene and its fundamental role in heredity is that mechanism.

A gene is the fundamental unit of hereditary information. Genes contain information in chemical form that when expressed can orchestrate, along with the environment, the development of an individual from fertilization to death. Since Darwin's time, modern genetic science has provided us with a greater understanding of genes. Gregor Johann Mendel is often called the "father of genetics," and although both Darwin and Mendel lived in the same era, it was not until recent times that we came to realize how much the field of genetics would enhance our understanding of the mechanisms to modify, store, and transmit hereditary information from cell to cell and from generation to generation. The information contained in a particular gene takes physical form as a unique sequence of four chemical bases on a strand of DNA. The chemical bases are adenine, cytosine, guanine, and thymine (A, C, G, and T).[6] The order of these bases in a gene serves as a set of directions for a cell to construct a protein (a gene product). Amino acids are universal building blocks for proteins, and specific gene sequences yield specific proteins. Some proteins are structural, such as keratin in fingernails; other proteins are enzymes that facilitate chemical reactions in an organism, such as the amylase digestive enzyme in saliva; and some are regulatory. Consider what happens when we eat a piece of chicken. First our digestive system disassembles the chicken proteins into the 20 amino acids that serve as building blocks for all kinds of proteins. Some of the amino acids are then broken down for energy upon delivery to our cells by the circulatory system, and others are reassembled by our cells to form human proteins as needed. The reassembled proteins contain amino acids in quantities and sequences that reflect the specific chemical base sequences (genes) in our DNA. Thus, we can recycle amino acids by turning chicken proteins (or any other kind of protein) into human proteins. Our genes carry all of the information to make this conversion.

Genetic information is coded in a chemical language that all forms of life can read and express. That is why a common bacterium such as E. coli, when given the human insulin gene, can produce human insulin for us to use. Genetic information is stable over time, yet capable of variation to form alternate gene forms (called alleles) and even entirely new genes. On the one hand, stability is necessary so that the DNA directions to build proteins can be relied on for survival generation after generation. On the other hand, a capacity for occasional variation in the DNA directions is essential for survival because natural selection favors new structures and processes that increase survival. New versions of existing genes may arise as variations accumulate. For example, human and bovine genes for insulin probably arose from an existing gene through changes (or mutations) in the genetic code. The human gene to produce insulin is slightly different than the bovine gene, yet the insulin produced by these two distinct genes is similar enough so that bovine insulin can be used to treat human diabetics. But because bovine insulin is not as effective for long-term treatment, technology developed to insert the human insulin gene into a bacterium for large-scale production of human insulin, which was quickly adopted when developed. Technologies to move genes continue to develop as geneticists explore therapy techniques. It is not easy to place a gene at the correct location for normal function in a cell. That is why we need to know more about where genes are located.

Genes are carried on chromosomes located within cells, and they contain all of the information necessary to build, to operate, and to deconstruct the cell. Cells are fundamental to life, and they play an essential role in the organization and transmission of genetic information from generation to generation to maintain a species. Some organisms such as the bacterium E. coli are single-celled and can only be seen with a microscope. Multicelled organisms are generally visible to the unaided eye and are made up of cells that, as they develop, take on special roles to form tissues, organs, organ systems, and the organism as a whole. Chromosomes are structures within cells that organize and carry the DNA. In organisms that contain cells with a nucleus, each strand of DNA is held in place with proteins to form a chromosome.[7] Human body cells, called somatic cells, have 23 pairs of chromosomes; typically they have 1 pair of sex-determining chromosomes and 22 pairs of autosomes (not sex-determining chromosomes). Each member in a pair of autosomal chromosomes is the same size and shape and has the same gene order. Sex-determining chromosomes come in two forms, X and Y. Females carry two X chromosomes in each of their cells. Males, however, carry an X and a Y chromosome in each of their cells. The X and Y chromosomes have few genes in common, and even though the Y chromosome is much smaller, a portion of the X and Y chromosome pair for distribution during sex-cell production. The Y chromosome is responsible for fetal development of the male by causing what would be ovarian tissue to descend and form testes. Two X chromosomes are important for normal female development, but only one X chromosome is essential thereafter, and thus in females one X chromosome randomly condenses to become inactive in each cell line. Since both males and females need one active X chromosome, it is easy to distinguish male and female cells under a microscope by looking for the condensed X chromosome in females that is not found in normal males. This condensed and inactive X chromosome is called a Barr body and influences the expression of genes on the X chromosome in females. For example, genes for some forms of hemophilia are carried on the X chromosome.[8] If a woman carries the disease-causing version of a gene for this disease on one of her X chromosomes but not on the other, then when her X chromosomes condense randomly during development, multiple cell lines form with each line then expressing genes on one or the other X chromosome thereafter (we call this a mosaic pattern of cells). The severity of her hemophilia is dependent on which tissue lines are affected. A man who carries the hemophilia gene on his X chromosome will have all cell lines affected and thus have a severe form of hemophilia.

We use the term homologous when referring to chromosome pairs that are the same size and shape.[9] Chromosome pairs are also homologous in that they carry genes, and other identified sequences of DNA, at the same location (called a locus) along the linear DNA molecule. Cells with pairs of chromosomes are called diploid, and this type of cell is typical for cells that make up the body. Cells with only one chromosome from each pair are called haploid, and this is typical for gametes (sperm and egg). Your diploid body cells were formed when your mother's haploid egg cell met your father's haploid sperm cell (fertilization) to form a diploid zygote cell—soon to be you. That zygote then became two cells, the two cells then became four cells, the four cells then became eight cells, and so on. With each cell division, the chromosomes and their DNA replicated, and then with a high degree of accuracy, they separated and moved to each descendent cell. This type of cell division conserves genetic information as it is doubled and then transmitted from cell to cell in development of a diploid organism from zygote to adult. However, a second type of cell division is necessary to create haploid gametes to begin the next generation. It is important to distinguish how genetic information is transmitted from cell to cell, and from generation to generation if we are to anticipate prospects for genetic interventions and future generations.

Two types of cell divisions distribute genetic information. First, mitosis is a type of cell division that yields new diploid body cells for growth following fertilization to form an adult. Mitosis is also involved in the ongoing repair of tissues. This type of cell division produces body cells that contain a complete genetic complement and all of the necessary fluid (cytoplasm) and parts (structures and organelles) for continuation of the cell line. These diploid cells make up the body, but they do not give rise to the next generation unless they are part of the germ-line tissue that includes gamete-producing cells in the testes and ovaries. Germ-line cells undergo a second type of cell division called meiosis to form haploid sperm or egg. Each sperm and egg cell carries the 23 chromosomes, unpaired, and all of the necessary cellular components for survival. Because males carry an X chromosome and the Y chromosome, they produce X sperm and Y sperm and therefore determine the sex of the child when a sperm unites with the X-carrying egg. Sperm are built for travel and therefore carry less material than eggs. Eggs are larger because they contain all of the materials necessary for development on fertilization. Nuclear DNA is not the only DNA replicated and distributed in cells with a nucleus. Mitochondria are membrane-bound structures (called organelles) that contain DNA and replicate independently for distribution during cell division. Mitochondria are sometimes called "powerhouses" because they are involved in energy conversions, and several genetic diseases are associated with this function of mitochondria.[10] Because mitochondria passes from mother to child via her egg, and the Y chromosome passes only from father to son, geneticists are able to trace the direct genetic history of the maternal and paternal lines, respectively.

Germ-line tissue is our past, present, and future. It is of great importance because it contains genetic information that was transmitted from previous generations and also contains genetic information that may be passed to future generations. At any given time, mutations add a source of variability—they might have no effect, or they might be deleterious, or they might be of potential benefit in the present or the future. Additional variability is introduced when a haploid sperm and a haploid egg are produced, and then recombined through fertilization to form the diploid organism.[11] The fact that both chromosomes and genes come in pairs is an important way for nature to ensure genetic diversity.[12] To clarify this point, it is necessary to examine the fact that somatic cells carry genes in pairs.

Just because our somatic cells carry genes in pairs does not mean the members of each pair are identical in DNA sequence. Consider the following analogy. Genes are like stories with an intended message for survival to be told over and over again from generation to generation. *The Boy Who Cried Wolf* story is retold generation after generation, and although the story may be modified to fit our circumstances as it is retold, the intended message remains the same—do not lie about danger for fun or to get attention! This is a message for survival that we teach our children. In a similar way, genes are replicated generation after generation with variation introduced into the genetic story by spontaneous or environmentally induced changes in the DNA base sequence (mutation). Just like the message in *The Boy Who Cried Wolf* story, the intended function of a gene is survival of the species. Changes or variations in the genetic messages are noticeable as alternative forms of a gene. Geneticists use the term allele when referring to these alternate forms of the same gene. It is possible for each member of a chromosome pair to have a different allele at a given gene locus. Thus, homologous pairs have the same gene order, but not necessarily the same alleles. Take a minute and touch your earlobe. Is it unattached or attached to the side of your neck? If your earlobe is attached, then you probably carry two copies of a recessive allele involved in development of the human earlobe. That does not mean that you are in some way inferior; it just means that you are not carrying at least one copy of the dominant allele that is responsible for development of an unattached earlobe. The terminology can be confusing. To say that a person carries a gene for attached earlobes and unattached earlobes is misleading because we are not talking about two genes; we are talking about two different alleles for the earlobe trait gene. The term allele has been used here when referring to a variation in the coding sequence of a gene for a trait, but the term is also used when referring to noncoding variations in DNA sequence. Single nucleotide polymorphism (SNP) is a term used in reference to a single base difference in DNA sequence. SNPs are found anywhere in the genome (between genes, or within genes in sequences that do or do not code for protein). SNPs are important tools for geneticists in their study and identification of genetic traits in individuals and populations.

Gene expression at the cellular level is the process by which protein is synthesized in accordance with directions contained within the genes, and those proteins then become part of processes to develop and maintain the organism. Sometimes we can understand how a gene works at the cellular level by examining a simple trait that is the result of a genetic mutation. For example, a mutation in the gene responsible for an enzyme called phenylalanine hydroxylase causes a disease called phenylketonurea (PKU).[13] Individuals who carry two mutated forms of this gene (two recessive alleles) will have the disease because they do not have a normal enzyme for the breakdown of a dietary amino acid called phenylalanine. Mental retardation is a result. Fortunately, a simple test can be used to identify babies with PKU disease, and subsequent symptoms are reduced by limiting phenylalanine in the diet. This is also an example of how environmental factors can play a role in the expression of a trait. In most cases, however, it can be said that traits are also polygenic; that is, many genes and their respective alleles are involved in the expression of a trait. For example, dozens of genes are necessary for normal hearing, so to say that there is a gene for hearing is inadequate and even misleading. These kinds of traits are complex because they involve many genes along with environmental factors. There are, however, well-known examples of less complex traits. A few are introduced here to create a framework to assist us in analyzing emerging genetic technologies.

Genes for autosomal traits are carried on the autosomal chromosomes (not the X or Y chromosome). A common example is the cystic fibrosis transmembrane conductance regulator (CFTR) gene located on chromosome 7.[14] This gene is responsible for a type of protein channel in cell membranes to connect a cell's interior and the fluid space between cells. Abnormal (or so-called recessive) forms of this gene are a cause of cystic fibrosis. If an individual carries two abnormal copies, they are said to be homozygous recessive for the CFTR gene, and their cells will produce a thick mucus instead of a thinner lubricant. Cystic fibrosis exhibits a wide range of symptoms such as lung congestion and subsequent infection.[15] If an individual carries a normal and an abnormal copy of the gene, they are said to be heterozygous or a carrier for the trait and their symptoms are negligible. We call the normal CFTR gene dominant because it masks the effect of the abnormal copy, which in this case allows at least some normal membrane channels to form. To be symptom-free, a person must carry two copies of the normal CFTR gene for this trait (homozygous dominant). Parents with cystic fibrosis in their family are frequently interested in understanding the probability of their child having cystic fibrosis, and to obtain this information they should seek the services of a genetic counselor.[16] We explore the possibilities here because traits such as cystic fibrosis are easily observed and have genes that distribute independently of one another during meiosis; such traits follow a simple Mendelian inheritance pattern.[17] Because cystic fibrosis involves a single gene on a single chromosome, we use this trait as an example of how we can estimate a child's risk of having the disease.

The following question and answer illustrate how an understanding of genetics can help us make reproductive choices. For example, what are the chances that two parents, both carriers for cystic fibrosis, will have a child with cystic fibrosis? Cystic fibrosis follows a simple Mendelian inheritance pattern. We could visit genenames.org for official gene/allele symbols.[18] But to keep things simple, we define our own symbols with uppercase C to represent the normal allele, and lowercase c to represent an abnormal allele. If both parents are carriers, then they each have a genetic makeup (or genotype) that can be represented as Cc. Recall that meiosis is a type of cell division that occurs in testes and ovaries to separate homologous chromosomes and their genes to form haploid sperm and egg. The father produces nearly equal numbers of sperm that carry C and sperm that carry c. The mother carries eggs with C and eggs with c in nearly equal numbers. Thus there are only four possible combinations for these sperm and egg at fertilization (CC, Cc, Cc and cc). Thus when both parents are carriers (Cc genotype) for cystic fibrosis, they have a one in four (or 25 percent) chance of having a child with cystic fibrosis (the cc genotype).[19] It is very important to remember that the birth of each child in this example is an event that is independent of previous births, and that a one in four chance of having a child with cystic fibrosis does not mean that after three normal children the next child will be affected. This simple method to predict autosomal traits can also be used when considering sex-linked traits.

Genes for sex-linked traits are carried on the X and Y chromosomes, but sometimes geneticists are more specific and identify a gene location as X linked or Y linked, respectively. Human females carry two X chromosomes, one inherited from each parent. In contrast, males receive their X from their mother and their Y from their father. Green colorblindness is an example of an X-linked recessive trait, and *OPN1MW* is the official name for the gene that in its normal form contributes to the production of a pigment in the retina that enables a person to see green colors, as most people do.[20] To identify traits as X or Y linked, we usually place the gene symbol as a superscript following the X or Y. The *OPN1MW* gene name is too large so it makes sense here to use a c for the recessive allele that causes color blindness and a C for the dominant version of the gene. X^C represents an X chromosome carrying a normal color vision allele, and X^c represents an X chromosome carrying a recessive allele. Because this is an X-linked trait, a gene is not identified on the Y chromosome. Consider the following genetic situation. Mary has normal vision but her father was color blind, so what are the chances that her sons will be color blind if she has children with Tom, who also has normal color vision? The first step is to establish the genotypes for Mary and Tom. Mary knows that her father was color-blind, and thus his genotype must have been X^cY. Mary received one of her X chromosomes from her father (in this example that X chromosome carries the colorblindness allele) and because Mary has normal color vision, her genotype must be X^CX^c. Her husband has normal vision and must have the genotype X^CY. The second step is to identify the types of gametes Mary and Tom carry with respect to the colorblindness allele. Mary carries X^C and X^c eggs (probably in nearly equal numbers). Tom produces X^C and Y sperm. The final step is to predict all possible combinations (genotypes) and their probabilities. The following genotypes are possible for their children: $X^C X^c$, $X^c X^c$, $X^C Y$, and $X^c Y$. Chances are one in two (or 50 percent) that a boy will be colorblind. All of the girls will have normal vision, but the chances are one in two (or 50 percent) that a girl will carry the colorblindness allele. We do not consider a Y-linked trait because it is easy to see that these traits always pass from father to son. Predicting genetic outcomes becomes more complicated when two or more genes are involved.

Multiple alleles frequently exist for a given gene locus. The ABO blood group system is a classic example of multiple alleles and how genes are expressed. The gene locus for this blood group system is on the long arm of chromosome 9 at band 34 (we use 9q34 as shorthand).[21] Bands are dark staining regions (visible with a microscope) on either side of the primary constriction to which the cell division apparatus attaches on a chromosome (called a centromere). The shorter of the two arms is designated as p and the longer arm as q; bands increase in number as the address (or locus) moves away from the centromere. Scientists have determined that in human populations we normally observe three variations in a gene at the 9q34 location that produces a surface protein (called an antigen) on red blood cells—the A, B, and O alleles. We carry two number 9 chromosomes, and can therefore only carry two of the three possible alleles in the following combinations (or genotypes) in our cells: AA; AO; BB; BO; AB; OO. Genotype is not always expressed as an observable trait, which is called a phenotype. The A and B alleles each produce a characteristic antigen that can be observed with a simple blood test; this is not the case for the O allele. Thus we say that in the ABO blood group system, the A and B alleles share dominance (they are codominant alleles) over the O allele (a recessive allele). The genotypic combinations in the ABO blood type system yield the following four phenotypes: type-A blood (AA, and AO genotypes), type-B blood (BB and BO genotypes), type-AB blood (genotype AB), and type-O blood (genotype OO). The ABO blood type system is a clear example of how multiple alleles at a gene locus work in combinations to yield several phenotypes, but it becomes more complicated when one or more gene loci are involved.

Multiple genes can interact to modify the expression of traits. The Bombay phenotype is a simple example of gene interaction. Individuals that exhibit this very rare phenotype will always test as having type O blood.[22] Normal expression of an ABO genotype (located at gene locus 9q34) is completely dependent on normal function of another gene (*H*) at locus 19q13.3 (on a different chromosome). The protein product of this second gene is necessary for the attachment of the A and B antigens to the red blood cell surface. If an individual has two defective alleles (*hh*) for the *H* gene, the A and B alleles are not expressed as a blood type. Thus the *hh* genotype causes a type O phenotype by overriding a person's ABO genotype.

The internal physiology of an organism and its external environment can affect gene expression. Internal cellular conditions influence genes, and in some cases as we will see when considering stem cell production, genes can be reprogrammed to yield other cell types. For our purposes here, sex-influenced traits serve as a simple example of how the internal physiology of an organism is a factor in gene expression. The gene for male pattern baldness (MPB), mapped to chromosome number 3, has been observed to cause early onset pattern baldness in men more frequently than in women. It was first assumed that the heterozygous condition was expressed as MPB in men but not in women; however these observations continue to be investigated.[23] This is a case where the same allele behaves differently depending on physiological conditions within the individual organism. Environmental factors external to the organism may also influence an organism's physiology to influence expression of its genes. For example, consumption of fava beans (a type of broad bean that looks like a larger version of a lima bean) causes hemolytic anemia (a rupture of red blood cells) in individuals with an abnormal X-linked gene. That gene is responsible for production of a red blood cell enzyme called glucose-6-phosphate dehydrogenase (G6PD).[24] Individuals carrying this abnormal gene do not show symptoms unless they consume fava beans. Thus, fava beans are the environmental trigger that causes this trait to be expressed. In this case, a single environmental factor was easily identified, but in a complex trait such as behavior, multiple environmental factors are involved and difficult to identify.

Multifactorial traits are complex traits that are influenced by environmental factors along with multiple genes (polygenic) and alleles. Traits such as height, skin color, and behavior are multifactorial and therefore exhibit a wide range of variation. A genetic predisposition for tallness depends on nutrition for full expression. Skin color is a polygenic trait with alleles at several different gene loci in many different combinations that result in a wide range of skin tones. Behavior is even more complex than skin color because of multiple environmental influences, and the extent to which genes are involved is poorly understood. For example, studies of families and twins have demonstrated an association between obsessive compulsive disorder (OCD) and SNPs within serotonin transporter and receptor genes. Millions of people are affected with obsessions and compulsions.[25] It remains to be seen to what extent complex traits such as OCD can be understood in terms of genes and the environment. Is OCD a disorder, or is it just a source of variation that may facilitate the survival of our species? Is OCD just another example of our genetic diversity? These questions can be raised about many so-called genetic abnormalities.

The genetic diversity of a species comprises past and present variations in genetic information within the germ-line. Mutations are a source of that variation. The fact that genes come in pairs is what makes it possible for some mutant genes to remain common in the gene pool. For example, the heterozygous genotype for cystic fibrosis offers some natural resistance to cholera. Cholera is a deadly bacterial disease that causes diarrhea, which leads to severe dehydration. The mechanism for this natural resistance is complex, but this phenotype is an example of a heterozygous advantage for those humans living in parts of the world where cholera is common. Thus the allele that causes cystic fibrosis remains in relatively high numbers in the human gene pool despite the fact that in a homozygous condition the disease is devastating. Genetic diversity is necessary if a species is going to survive epidemics such as cholera, and also for a species to adapt when its environment changes. Recall Darwin's finches. Natural variation (genetic diversity) in finch beaks enabled some birds to survive by accessing different kinds of seeds when food became scarce; this natural variation played an essential role in survival of the first finch population on the Galapagos Islands. Thus genetic diversity is like a savings account of back-up genes and alleles for a rainy day. Genes and their alleles can remain stable in a species, yet they are capable of variation through mutation and natural selection. It is this delicate relationship that natural selection acts on to program a species for survival in a particular niche. We should keep this delicate balance in mind as we genetically modify ourselves and other species on earth.

Humans have been influencing the genomes of other species for a long time. The evolution of modern-day bread wheat is believed to be a direct result of human intervention and several mutations. Early grasses are believed to have crossed naturally with other species of grass to yield sterile hybrids, but mutations occasionally occurred in grasses to allow fertile varieties through chromosome doubling. This then allowed normal sorting of the genetic material during gamete formation to form grass hybrids with viable seed. This is a normal process in plants and can result in robust plants and larger seed. Our human ancestors may have encouraged development of modern-day bread wheat through the practice of selecting the largest seeds for planting the next generation.[26] Thus our ancestors and their agriculture evolved together (not unlike the coevolution of flowers and pollinators). So it can be said that through animal and plant breeding, our ancestors did in fact genetically modify organisms, but they did so within environmental constraints—they could not make the mule fertile.

Today, we can alter the genetic make-up of organisms. But genetic manipulations are not like working on a mechanical device such as an engine that can be turned off if it goes haywire; we might not be able to stop this "engine." Biological organisms have the ability to expand in space and time, and even as we take steps to prevent reproduction of modified organisms, there is no guarantee that genetic interventions will avoid the germ-line. Recall that germ-line tissues are a species' past, present, and future. Genes passed from generation to generation endow each species with a program for survival in a particular niche. We should constantly remind ourselves that it took a great deal of time for these complex programs to develop as each species adapted to function in its specific place in our global ecosystem. Should we be manipulating this fundamental unit of evolution and life as we know it?

MORAL REASONING

To approach issues raised by the new genetic science and technologies, it is important that we consider how best to approach them by using some of the basic tools of moral reasoning. In this section, we offer some general information on ethics, definitions and important distinctions in ethics, different ethical theories, and other noteworthy concepts that we need to know as we move forward.

Virtually since the origins of human history, humans have seriously entertained the question of how to be a good person and how to live a good life. This has been a universal quest of humans, whether religious or not, and it has certainly been a focus within philosophical and religious traditions. Specifically with reference to the three fundamental questions addressed by all religions, morality focuses not on the questions of why we are here or where we are going but rather on how we should live. It is important to note that the branch of study known as ethics has both secular and religious components, and we need to be acquainted with both; in fact, there is considerable overlap.

One of the foundational questions with regard to morality has to do with where it comes from. Is it rooted in human nature, is it rooted in God, or is it rooted in biology? The answer to this question depends on one's perspective. Most philosophers would agree that our understanding of morality is or at least should be rooted in reason, which is considered a key component of our human nature. In fact, one could say that reason is the principal source for a secular approach to moral questions. Those in religious traditions, on the other hand, would say that in some way, morality is rooted in God—it derives from God, is made known by God, and reflects the character of God. Some biologists, however, especially those who work in the field of sociobiology, would say that what we typically refer to as moral behavior is actually rooted in biology. This apparent moral behavior is related to that which is part of our physical being, including that which we have in common with other animals, and that which is a part of our evolutionary history and evolutionary continuance as a species.[27] For example, a biologist might argue that the self-sacrifice exhibited by mothers toward their children can be explained as a maternal instinct common to all species. In addition to the biologists, philosophers, and theologians, though, there are individuals who say that morality is something created by individuals as a result of their free choice (existentialists) or that morality is the result of an agreement that humans in society make with regard to rules of behavior that would best benefit everyone overall (social contract). This is all to say that there are many different ways of conceptualizing the origins of morality, although in some cases, the answer to what is the moral thing to do in a particular situation can sometimes be virtually universal (e.g., not molesting children).

Because evolutionary theory is such a fundamental part of biology, and in fact is the major paradigm within which biologists work, it is necessary to briefly address this in light of theology. Before Darwin as well as before the Enlightenment, virtually all Christians believed that the world was created as described in the early chapters of the book of Genesis—God created the world out of nothing; God created a literal man, Adam, and a literal woman, Eve, who were the first humans and parents of us all; all animal species were created in their fixed forms; humans had dominion over all other creatures; and the entire world was believed to be only several thousands of years old. The rise of historical criticism, in which the Bible was analyzed in a way similar to other ancient texts, raised questions about how literally certain portions of the Bible should be taken, and certainly Darwin's theory undermined not only the entire creation account but also the crucial central role that humans play in God's plan. In the modern world, Christian believers are divided in terms of how they think the theory of evolution can align with their faith tradition, if at all. Most mainstream and liberal Christian denominations and individuals, as well as the Catholic Church, have no problem accepting the theory of evolution, as long as it posits the existence of a God who originated the process (and for some, who continues to work within nature). Most conservative Christian denominations and individuals reject the theory of evolution as incompatible with the Bible's account, and many of them maintain that "creationism" be taught alongside of evolutionary theory as an alternative theory within the biology classes of public schools. In any case, two important points need to be made. First, belief in evolutionary theory is not necessarily at odds with Christian faith. Second, regardless of one's belief in or rejection of evolutionary theory, it is not crucial to considering the important genetic issues in this book.

It is important to define what is meant by the terms ethics and morality. These terms are often used interchangeably by most of us, such that when we say that someone is an ethical person or a moral person we are essentially saying the same thing. The terms are often distinguished by ethicists, however, and the distinctions are important. Ethics can be defined as systematic reflection on the question of the right and the good.[28] Morality can be defined as the living out of this reflection, such as in the actual decisions that we make. Ethics is more of an academic exercise in which we stand back from an issue and analyze it from various perspectives, using our capacity to reason. Morality is what we do as a result of our reasoning. Let us take the concrete example of abortion. A strictly ethical approach would examine the issue of abortion in terms of the arguments made by the pro-choice side (that abortion should be permitted since women should have autonomy over their bodies) and those made by the pro-life side (that the embryo is human life and should be protected at virtually all costs). An ethical approach would also want to understand not only the arguments and the reasons why people hold their positions, but also the deeper points over which they disagree and which divide them into these different camps. Some important points in the abortion debate would be the moral status of the embryo: Whose rights should prevail when they are in conflict—the mother's or the embryo's? When does human life begin? Is there a difference between human life and personhood? What about the woman who finds herself in the situation of having an unexpected pregnancy, though? She is in the position where she must make a decision. She will probably consider some of the ethical arguments on abortion but may in fact already know what she will do because of her previously established views on the issue. But ultimately, she will have to decide what she herself should do in this situation.

What this example demonstrates, though, is that there is and should be a relationship between ethics and morality. We certainly should reflect long and hard on difficult issues, and at other times we will need to make decisions without the luxury of extended ethical analysis. With regard to this book and the issues we are exploring, though, calling this section "moral reasoning" is our attempt to bridge ethics and morality. In fact, we usually use the terms ethics and morality interchangeably. But the focus of our ethical discussions is not so much on what any one individual should do when faced with a choice, for example, about whether they should select for the sex of their child, but rather to look broadly at all sides of the issue, with the resultant goal that it will certainly help guide individuals in their actual moral decision-making. One of the major goals of ethical analysis is not necessarily to change one's mind about issues, or to reinforce one's already deeply held convictions, as much as to be as informed as possible about all of the arguments, positions, and issues within the issues as possible. It will certainly mean that not everyone will come to the same conclusions.

Ethics is also a technical branch of study within the disciplines of philosophy and theology.[29] Much of philosophy is directed toward understanding the moral life in general, particularly by studying some of the great philosophical ethicists in human history, including going back to Plato and Aristotle.[30] The branch of philosophy that studies particular moral problems is called applied philosophy. Thus, considering whether or not we should engage in genetic engineering would fall under the purview of applied philosophy. Within the Christian tradition, all study of the moral life is referred to as ethics within Protestantism and as moral theology within Catholicism. The terms are basically synonymous, but what is especially helpful about the Catholic terminology is that it draws attention to the fact that Christian ethics is and must be rooted in theology.

The study of ethics in both secular and religious traditions has similar goals: to understand the nature of human beings; to consider concepts such as human reasoning and human freedom; to address how we actually engage in moral reasoning and to argue for one method as opposed to another; and to determine whether or not moral absolutes are possible and if so how we can know these moral absolutes. Ethics has both a subjective and an objective side; that is, there are subjects of moral action as well as objects of moral action. Within the ethical tradition, most have believed that only humans can be subjects of moral action, who can actually engage in reasoning about different possible courses of ethical behavior. Both humans and other animals, however, can be objects in that the moral actions of humans can affect them for better or worse. Thus, if a human engages in what we would usually consider the unethical practice of lying, it hurts other humans, and that is part of what makes it wrong; if the same human gives a gift to another, that is usually considered a good thing to do. Similarly, if a human being tortures a dog, we would say that the human is engaging in an unethical behavior, partly because he or she is harming the dog; if the same human gives a dog a treat, we think he or she is doing something good. Most humans do not think animals capable of making the same kinds of moral choices as humans, and that when animals do things that seem to be moral or unselfish (such as dolphins rescuing drowning swimmers), we think that there must be some instinctual basis for this behavior. It is important to note, though, that there are many individuals who study the ethical treatment of animals, and those who actually study animal behavior and cognition, who believe that at least some animals are capable of truly moral behavior.

The issue between subjective and objective morality also affects the question of who or what constitutes a being that should be worthy of moral consideration. For example, an important distinction is often made between humans and persons in medical ethics. Not everyone agrees with this distinction, and some religious people, in fact, find it quite troubling. It is, however, important to mention for many of the genetic issues to follow, especially with regard to any technologies that deal with embryos. To be a human is to be one who has the human DNA and human genome, in essence, to be one who belongs to the species *Homo sapiens*, who has been born to human parents. To be human is in essence, then, rooted in one's biology. To be a person, however, is usually associated with certain cognitive abilities such as ability to reason, ability to have social interactions, ability to postulate a future, ability to plan goals. If one holds to this schema, then there are four possible combinations. One can be a human and a person (this includes all cognitively capable adult humans); one can be human but not a person (this includes those with cognitive deficiencies, including embryos, those with advanced dementia, those who are severely mentally challenged, those in a comatose state); one can be a person and not a human (some argue that there are animals with cognitive abilities greater than that of at least some humans); and one can be neither a human nor a person (includes most if not all animals and certainly plants). The significance of personhood usually has to do with which beings have rights and to what extent the rights of some beings should trump those of others, as well as the idea that when there is a limited amount of distributive goods (such as in medical care), that preference be given to persons. In any case, this is a controversial concept for many religious people, but is a very common distinction in medical ethics.

Finally, one other important point to make about definitions of ethics is that there are different ways of conceptualizing ethics. We have already distinguished between philosophical and religious ethics. Both secular and religious ethical systems, for example, distinguish among areas of focus, such as between personal and social ethics; the former focuses more on individual human morality and the latter on issues related to the social good. However, the two are also related. Some of the areas of social ethics include political ethics, medical and biomedical ethics, environmental ethics, economic ethics, ethical treatment of animals, and ethics of war and peace. Of course, individual decisions are made within all of these areas, so that even if a country or a society or the world community decides the parameters of when war should be undertaken, a personal decision is still needed about whether a particular individual drafted to go to war will actually fight. Another distinction has to do with the discipline within which the ethical discussion is taking place, as well as the subject matter, so that we can talk about business ethics, legal ethics, and pastoral ethics, for example. In addition, ethics can be divided into different approaches to ethics, which are usually determined by the sources they use. For example, liberationist approaches in ethics take as their starting point the experience of a particular oppressed group; thus feminist ethics takes as its starting point the experience of women with regard to their inferior status throughout most of human history. The ethical issues that new genetic science and technology raise fit into the category of social ethics, but have implications for personal morality as well. Thus, we need to consider as a community what kinds of genetic screening for defects is a good thing, for example; but if the technology becomes widely available, individuals will have to make personal decisions about whether they want their embryos screened for defects as well as the action they will take if that turns out to be the case.

An important component of ethics has to do with ethical theories, both secular and religious ones. There are many different kinds of theories, but we discuss only the more common ones. It might be helpful to think, as you read the descriptions of the theories that follow, about which is your typical way of approaching ethical issues. Often ethical debates occur when people are talking past each other on the basis of different ethical theories. The best way to explore and better understand the basic approaches of each of the theories is to know what aspect of human action it focuses on. Any moral act, or action, really consists of three different components—the person who does the action, the action that the person undertakes, and the effect that the person's action has on others. This is true whether reasoning from a religious or philosophical approach. Let us consider the following examples of moral or immoral actions. Person A takes care of Person B's children for the afternoon to help B out. Person C tells a lie to Person D which hurts D. Whether a good or a bad deed is done, someone is doing something, something is being done, and someone is having something done to them. There are different ways to conceptualize these three components; one Christian text uses the terms character, choices, and community. Other corresponding terms one can use are agent, action, and consequences. No matter which terms are used, though, they all correspond to this threefold distinction.

Many maintain that all three components are important in assessing moral behavior, but individuals also tend to emphasize one more than the others. Let us take another example: you come across a person holding a smoking gun in his hand and another person lying on the ground with a bullet hole in her. Let us assume for the moment that we can be sure that the person holding the gun is indeed the one who shot the other individual. The ethical question is: Did he or she do something morally wrong? Well, does this not depend on the full circumstances of the situation? We can assess this situation by exploring several different scenarios. If the person with the gun was someone who premeditatedly killed the other out of hatred, we would probably say that was wrong. But what if the person was a state-appointed executioner shooting a person on death row, or a soldier killing an enemy soldier, or a person shooting in self-defense because someone had threatened his own life? We would probably assess it differently. In the original example it was indicated that the person was shot. Does it make a difference if the person were only wounded or if she were killed? Again, it depends. Even in our law we make a distinction between actual murder and attempted murder; thus a person who succeeds in killing another and one who doesn't succeed in killing another, but intended to, get different prison sentences. Whether these two persons are really morally different, though, may also depend on whether you think that it does not matter what the consequence is, but rather what the intention of the person was. Thus, we cannot necessarily say that shooting someone is always wrong, but there are certainly some actions that many would agree are wrong regardless of the intention or of the consequences (e.g., many people believe that abortion or euthanasia or even genetically engineering children is morally wrong, no matter what). This then brings us to a brief consideration of three categories of ethical theories.

Virtue ethics refers to those theories that focus on the agent/character/person engaging in (or about to engage in) the action. The focus here is on virtues, motivation, and intention of the agent. Virtue theorists maintain that if humans cultivate the right kind of character, then the right moral action should necessarily follow. Virtue theory was made popular by the ancient Greeks, with Plato and Aristotle being especially influential figures. Aristotle's thought became infused into Catholicism, in particular through the writings of Thomas Aquinas. All virtue theorists believe that the cultivation and practice of virtues, and the avoidance of the practice of vices, is essential to the moral life. Virtue and vice are usually defined by their habitual nature: to be a virtuous person is to be in the habit of practicing virtues. Examples of virtues might include honesty, generosity, and courage, to name a few. To have the individual virtue of honesty, then, requires that one tell the truth virtually always and not just occasionally. The vices work essentially the same way. Aristotle developed an interesting way of determining virtues and vices, in which he defined virtue as the mean between two extremes. Thus, generosity would be the mean between the extremes (and thereby the vices of) being cheap and being extravagant—"everything in moderation" would be Aristotle's maxim. Not everyone agrees with Aristotle, of course, and though virtue theories decreased in popularity particularly during the Enlightenment, they recently experienced a resurgence of interest, both in philosophical and theological circles.[31] Virtues are important to both secular and religious traditions. Virtues have even been divided into different categories: the cardinal virtues of prudence, justice, fortitude, and temperance,[32] and the theological virtues of faith, hope, and love.

In addition to the concept of virtue, these theories emphasize the intention and motivation of the agent. These concepts are often used interchangeably, but they do mean different things. Intention has to do with the goal that one hopes to achieve, and the motivation has to do with the reasons why one decides to do a certain action. Let us refer to the shooting example again. If the person with the gun wants to shoot someone because he wants to take revenge on someone who has harmed him, then we would say that revenge is the motive. If he desired to kill the person rather than simply injuring her, then we say that his intent was to kill rather than to injure. Motivation is especially important in criminal law, such that we take the motives (as best we can ascertain) of an individual into account when assigning them culpability. Taken together, then, virtues, intention, and motivation are at the heart of virtue theories of ethics.

To take an example from the new genetics, then, it might make a difference to us to know, for example, whether a particular researcher working on stem-cell research was primarily motivated by whether or not she would make a name for herself in the field as opposed to doing it to help better the condition of humankind. Of course, it is possible and not necessarily wrong to have multiple, or mixed, motives. It might make a difference to us what reasons a parent might have for wanting to genetically engineer their child— whether it was to screen for diseases or select for a particular sex. Virtue theories have been commended for emphasizing that because individuals make decisions, it is certainly important to focus on their character development. In addition, they confirm what most of us already believe: that what we do is related to who we are. However, virtue theories have been critiqued; one criticism is that "being" does not of necessity lead to "doing." In addition, it is difficult for humans to agree even on a list of what virtues should be cultivated; our lists would most likely be quite different from Aristotle's.

Deontological ethics refers to those theories that focus on the choice/action/decision itself. Deontological theorists believe that there are universal rules of morality which humans can discern, and that some actions are intrinsically wrong by their very nature. People who operate within these theories may disagree, however, about what they consider these intrinsically wrong actions to be. The intention of the agent or the supposed good consequences do not enter in. Deontology focuses on the concept of duty. The secular version of deontological ethics was developed by Immanuel Kant, and the religious version is referred to as divine command theory. One can uphold divine command theory but also consider Kant's theory to be helpful as well.

Immanuel Kant is considered the greatest modern philosopher, although some of his works are considered confusing even by scholars devoted to studying his ideas.[33] He believed that through human reason we can arrive at moral principles, or rules, that are universally binding. The motive, if one can indeed talk of a motive in deontological theories, is that of duty; for Kant, it is a higher moral good to do an action out of a sense of duty than out of an emotional response, for example. He is most famous for the "categorical imperative." It actually has several different forms, which has led some scholars to conclude that multiple imperatives exist rather than different versions of the same one. Regardless, the essence of the imperative is to assist humans in developing universal moral guidelines for behavior. The most popular version of the categorical imperative has to do with the concept of universalizability: before we engage in an action, we should consider whether or not we would want others to engage in this same action. This approach led Kant to the conclusion, for example, that actions such as lying and stealing were always wrong.

A version of divine command theory appears in virtually all religious traditions. It is similar to Kant's theory in that the focus is on duty, only in this case it is duty to God. It is also similar in that it maintains that there are universal moral norms. In response to the question in the mouth of Socrates in Plato's *Euthyphro*, "Is that which is holy loved by the gods, or is it holy because it is loved by the gods?", most religious traditions would maintain that somehow morality is derived from God. Divine command theory assumes the existence of a God who is moral, who desires that humans live moral lives, and who provides guidance to humans about how they should live. There are some obvious strengths with regard to this approach to ethics. Most humans have believed in some kind of god or gods, and so the idea that this god (or gods) has preferences for how one should live is plausible. In addition, religious traditions generally agree on what constitutes a moral life, at least in the broad strokes. For example, virtually every major world religion has some version of the Golden Rule: "Do unto others as you would have others do unto you." There are, however, problems with this approach to ethics. It is not much help for those who do not believe in God. There is the further problem of how can we really know what God wants. In addition, despite the similarity in broad strokes for ethics, religious traditions differ considerably, even those belonging to the same "family" (e.g., the various Christian denominations), in terms of what is right and wrong.

How might Christians derive morality from their faith in light of divine command theory? Here we must return again to the different sources that are used in Christian theology, which are also the same sources used in ethics: Scripture, tradition, reason, and experience. In some cases, it is obvious what "the" Christian answer to an ethical problem might be, and in other cases it might not be. Let us take the example of premeditated murder. If we examine this in light of these sources, we would see that the sources are in basic agreement: the Bible condemns murder; the Christian tradition has held that intentionally killing an innocent person is wrong; if we reason about it we see what the problems are with murder as a course of action; and experience confirms this. But what if we look at the example of whether war can be morally justified? The Christian tradition is still today divided on this issue. Even if we do not look at the other sources, the Scripture itself provides contradictory evidence; pacifists point to Jesus' teaching on "turning the other cheek" as an admonition to avoid violence, and just war theorists point to the prevalence of war in the Scriptures seemingly sanctioned by God and sinful human nature as pointing to the inevitability of war as sometimes being a necessary evil. And how do we even begin to think "Christianly" about issues that have no real history in our tradition and which are not in the Bible, such as to what extent should we modify genes to prevent or mitigate genetic maladies? There are general principles in the Bible in terms of the dignity of persons, and Jesus' mission of healing (also granted to his disciples), but where do we draw the line in light of new genetic technologies? Hence, specific guidelines on controversial areas of ethics are not easy for Christians to come to universal answers on, and some would argue that this is not even possible, although possibly desirous.

Consequentialist ethics refers to those theories that focus on the community/consequences/effects of the action itself. There are a few different versions of this, specifically utilitarianism and situation ethics (the former is often divided into two types). This type of moral reasoning requires a balancing of costs and benefits, with the recognition that there may be both good and bad consequences for the same action. Also, it is obvious that we cannot predict outcomes with certainty, so that the best we can do when making decisions under these models is to make an educated guess on the basis of intended, predicted, or expected consequences. For those adhering to a strict consequentialist ethic, this is really the only important consideration. The motivation for the action on the part of the agent and the action itself are secondary considerations, if they are considered at all.

Of the two basic versions of consequentialist ethics, the more common one is utilitarianism, which was developed by the philosophers John Stuart Mill and Jeremy Bentham. The emphasis in utilitarianism is generally on taking the action that you believe will bring about the greatest good for the greatest number; again, the focus is on expected or anticipated results because we cannot predict results with certainty. Interestingly, this theory includes the costs and benefits to animals as well as humans, so that all are included in the ethical calculus. It is important to note that there are varieties of utilitarianism; for example, act utilitarianism focuses on the consequences of individual actions, whereas rule utilitarianism maintains that some rules should be established which we know will bring about better consequences overall even if not with regard to a particular action. The latter has been criticized as really being a kind of deontological ethic and not a utilitarian one. A utilitarian calculation is almost always a necessity on the public policy level of decision making because policy makers, including government leaders, must or at least should take into account what will benefit most of the populace. Of course, there are exceptions, and in reality it does not always work because even policy makers can be self-interested. Also, what is good for the majority may not be good for the minority, whose interests also certainly count.

Situation ethics was developed by the Episcopal priest, Joseph Fletcher, and never quite caught on the same way as did utilitarianism.[34] He maintained that individuals should engage in those actions that they believe will bring about the most loving consequences. He focused especially on the Christian notion of agapé, which is the highest kind of love, expressed perfectly by God but only approximately by humans.

Consequentialist ethical theories certainly have their benefits. They emphasize the importance of the effects of human actions on others, something that would resonate with all of us. We would probably agree that public policy decisions in particular should not benefit just the few. But these theories are also prone to some problems; with both of them, there is the difficulty of not being able to predict consequences with certainty, and the danger of harmful and unexpected consequences. Utilitarianism is also problematic because it has the potential to significantly deny the rights of individuals if it will benefit the group. There is the additional problem of how to engage and who will engage in this cost-benefit analysis. Situation ethics is problematic because it leaves us with no universal norms and an absence of real content to the concept of love; and it ultimately results in actions that are uniquely individual, are relativistic, and ultimately have little need of justification. It is important to note that much of the ethical debate on the new genetics is often justified on utilitarian grounds (that it will benefit humankind as a whole). It is often criticized, however, both on utilitarian grounds and on deontological grounds. A utilitarian criticism might be that it will not benefit and may in fact harm humankind as a whole. A deontological criticism might be that there are some actions within the field of genetics which should never be undertaken, regardless of the possible good consequences, such as experimenting on embryos.

So, you may ask, how will we arrive at ethical answers to difficult questions when we factor in all of the above? It is not easy! The more complex and multifaceted the ethical issues, such as in genetics, the more difficult it is to come up with "the" answer, especially one that everyone can agree on. Every ethical theory has its own problems, but that does not mean that we need to completely disregard them. Just because Scripture does not address specifically the question of human cloning does not mean that there are not resources within the Christian tradition that can help us fashion an answer. We need to draw on the multiple, rich resources that have existed for so long within the discipline of ethics and use them as an aid to navigate the new issues with which we are confronted.

Finally, a few important concepts in ethics need to be briefly mentioned. First, it is important to distinguish between a universal and a relative approach to ethics. The former assumes that there are general moral norms or rules that we can agree on and according to which human societies should adhere. Religious traditions and any deontological ethical system will maintain this. A relative approach to ethics, on the other hand, maintains that there are no universal moral norms and that what is the moral thing to do is that which is decided either by individuals (such as in situation ethics) or by cultures, with the idea that what one culture may consider to be immoral another culture may consider to be moral. With regard to issues in genetics, using sex selection as an example, some will argue that the choice should be left up to individuals and couples, and that what is right for one is not necessarily right for another. Second, the concepts of the moral and the legal are distinct but related areas with regard to ethics. Often morality leads to legality, so that because we consider murder to be morally wrong, we enforce laws about it that will help ensure that it is not likely to be committed. Likewise, some laws prohibit certain behavior that we would not necessarily consider immoral, such as jaywalking. What is notewor-thy, though, is that when we consider something to be a serious moral issue, we often support it with legislation or governmental policy. For example, the federal government bans the use of federal funding for human cloning because most people consider it morally problematic.

Third, a common argument used on complex ethical issues is called the "slippery slope" argument; it is used often in the field of medical ethics. What the slippery slope argument maintains is that because there is usually a point beyond which we do not want to venture ethically, we may not want to approve of an action that could lead past that point. Let us look at an example in medical ethics: physician-assisted suicide. Although many are opposed to this practice, many are in favor of it under certain circumstances, such as when a person is terminally ill, is in extreme pain, is unlikely to recover, and wants to die. But even those sympathetic to this position might argue that legalizing this practice might move us down the slippery slope to where we eventually will allow physicians to help individuals die who are not in this situation. A counter to the slippery slope argument is the "wedge" argument, which main-tains that we need to simply be careful to insert specific guidelines that will prevent us from "sliding down." The slippery slope argu-ment is certainly an important one with regard to genetic technolo-gies. Thus, although many people might agree that genetically screening embryos for genetic maladies could be permissible, they might not be comfortable with enhancing embryos with nonmedical characteristics such as intelligence or athletic ability. Fourth, the issue of human rights and the corresponding principle of justice are important concepts in ethics, especially within medical ethics. To say that someone has a right to something is to say that the rest of the moral community has a duty to respect that right and help enforce protection of that right. Within medical ethics there are even Patients' Bills of Rights distributed to hospital patients so that they can know what kind of treatment they deserve and should expect. But the new genetics raises even new questions about rights. Do individuals or couples have a right to have a child technologi-cally if they cannot have one naturally? Do parents have a right to design their children? These are not always easy questions to answer. In addition, there are often situations in medical ethics where there are conflicts of rights, which raises the additional problem of how to mitigate these conflicts. The principle of justice is examined in a chapter of its own, but in essence, has to do with the concept of fairness. With regard to the new genetic technologies, an important justice question is who will have access to them—only the rich or all individuals who can benefit from them? And how will we ensure fair distribution? Finally, within the medical field are the principles of health care ethics, which are also used to guide ethical behavior of medical professionals toward their patients. The four that are universally agreed on are beneficence (doing good to the patient), nonmaleficence (not harming the patient), autonomy (cognitively capable individuals should be able to make their own decisions), and justice.[35]

CONCLUSION

Obviously, both genetic science and moral reasoning are very large areas of knowledge and inquiry. It is truly impossible to do justice to both of them in just one chapter, and so it is the foundational and conceptual on which we focused. To navigate the tough issues which follow, however, it is important to grasp the basic concepts of both areas. What we have focused on here informs the chapters that follow, but is expanded as well as we consider both the specific science and genetic technologies behind each of the five controversial areas discussed in the next chapters. We also need to consider the important moral and religious concepts that will be challenged by the new genetics, and perhaps need to be modified, as we deal with current and future technologies for which we do not have an adequate blueprint. But as Dr. Van Helsing says, because the human impulse is to continue guessing, we must ensure that it is educated, thoughtful, and ethical guessing that we engage in. Although we may not arrive at answers, we can at least move in that direction, recognizing that answers are not easy, that not everyone will agree on what the answers are, and that even if there is consensus on some of the answers, we are still left with the troubling prospect of instituting public policy. This policy ideally should truly protect the integrity of the human and other genomes, move beyond the immediate and think about the implications for future generations, and ensure that our journey, even if not our destination, is one marked by careful navigation and not simply aimless meanderings.

NOTES

1. 1 Corinthians 13:12.

2. Darwin was also influenced by the ideas of Thomas Robert Malthus, a political economist, who introduced the idea that as human populations grow, resources become limited. Competition for the remaining resources then increases and may lead to epidemics, war, and famine.

3. On very rare occasions (one in a million), a donkey stallion and a horse mare will produce a fertile mule mare. For additional information, visit the American Donkey and Mule Society at http://www.lovelongears. com/index.html (accessed March 16, 2008).

4. The term "meme" was coined by Richard Dawkins to describe cultural phenomena such as slogans, religions, fables, and so on that can pass from individual to individual and from generation to generation. Today evolutionary theory is used as a framework for analysis of both genes and memes. For example, questions are often raised about to what extent genes determine memes, and how do genes and memes interact to permit our survival.

5. Darwin's theory has been extremely influential, and unfortunately, it has therefore been misapplied to human societies. Darwin's use of the term "fittest" applies to an ability to pass genes into the next generation. This reproductive success does not necessarily require a fight for survival, nor does it imply that those who are less fit should intentionally be denied the opportunity to produce children.

6. Ribonucleic acid (RNA) is another information-carrying molecule. Information contained in the coded sequence on RNA is transcribed from DNA with the same letters except U (uracil) replaces T (thymine). The information transcribed to RNA is then translated by cellular systems to construct proteins.

7. In cells with a nucleus, each chromosome contains one complete strand of DNA that is wound on spool-shaped proteins for packaging into chromosomes that are easily visible with a microscope at certain times in the life cycle of a cell. In cells without a nucleus, such as bacteria, the DNA takes on a circular form without protein support and is readily available to be accessed throughout the life cycle of the cell.

8. http://www.ncbi.nlm.nih.gov/sites/entrez (accessed August 28, 2008).

9. A photograph of each chromosome pair in order from long to short is called a karyotype; for examples see http://www.genome.gov/ Pages/Hyperion/DIR/VIP/Glossary/Illustration/karyotype.cfm?key=karyotype (accessed August 12, 2008).

10. An excellent source of information on mitochondrial diseases is the United Mitochondrial Disease Foundation at http://www.umdf.org/site/ c.dnJEKLNqFoG/b.3041929 (accessed August 20, 2008).

11. Some sources of variability in meiosis include the exchange of genetic material between homologous chromosome pairs during meiosis, followed by random assortment of the homologous chromosome pairs to yield gametes with one member of each pair.

12. The Y chromosome is an exception. It contains very few genes in addition to those responsible for development of the male. The X and Y chromosome have very few genes in common.

13. The NCBI sponsors OMIM. This is an excellent Web site for information about known genes and associated traits. Regarding PKU, see http://www.ncbi.nlm.nih.gov/entrez/dispomim.cgi?id=261600 (accessed August 16, 2008).

14. OMIM at http://www.ncbi.nlm.nih.gov/entrez/dispomim.cgi?id= 602421 (accessed April 21, 2008).

15. MayoClinic.com at http://www.mayoclinic.com/health/cystic-fibrosis/ DS00287/DSECTION=2 (accessed April 21, 2008).

16. For more information, visit the Web site for the National Society of Genetic Counselors at http://www.nsgc.org (accessed August 15, 2008).

17. If two genes are on the same chromosome pair, they are called linked genes and they do not sort independently during meiosis to form gametes. Linked genes are, however, shuffled during an early stage in meiosis when homologous chromosomes pair tightly and exchange genes. Linked genes do not follow a simple Mendelian inheritance pattern.

18. Current gene symbols are readily available at HUGO Gene Nomenclature Committee http://www.genenames.org, sponsored by the National Human Genome Research Institute (NHGRI), accessed April 22, 2008. As new genes are discovered, criteria for naming those genes are important. Criteria can be found at http://www.genenames.org/guidelines.html#2. %20Gene%20symbols (accessed April 22, 2008).

19. The chances for birth of normal child (the CC genotype) are one in four (25 percent). The chances for birth of a child that is a carrier (the Cc genotype) are two in four (or 50 percent).

20. http://www.genenames.org/data/hgnc_data.php?hgnc_id=4206 (accessed April 22, 2008).

21. The ABO blood group is one of the best known examples for illustrating basic genetic concepts. For additional information, visit OMIM at http://www.ncbi.nlm.nih.gov/entrez/dispomim.cgi?id=110300 (accessed April 13, 2008).

22. Additional information on the Bombay phenotype is available at http://www.ncbi.nlm.nih.gov/entrez/dispomim.cgi?id=211100 (accessed April 15, 2008).

23. OMIM at http://www.ncbi.nlm.nih.gov/entrez/dispomim.cgi?id=109200 (accessed April 26, 2008).

24. OMIM at http://www.ncbi.nlm.nih.gov/entrez/dispomim.cgi?id=134700 (accessed April 26, 2008).

25. Serotonin is a neurotransmitter. For more information, see OMIM at http://www.ncbi.nlm.nih.gov/entrez/dispomim.cgi?id=164230 (accessed April 26, 2008).

26. Jacob Bronowski, *The Ascent of Man* (Boston and Toronto: Little, Brown, 1973), 64–68.

27. For a good volume on a Christian approach to biological origins, see Stephen R. L. Clark, *Biology and Christian Ethics* (Cambridge: Cambridge University Press, 2000).

28. A classic resource for the distinction between the right and the good can be found in W. D. Ross, *The Right and the Good* (Indianapolis and Cambridge: Hackett Publishing, 1930).

29. One of the best short but thorough introductions to ethics (particularly helpful with regard to ethical theories) is found in William K. Frankena, *Ethics*, 2nd ed. (Englewood Cliffs, NJ: Prentice-Hall, 1973). James Rachels also wrote helpful textbooks on moral philosophy which are highly readable and became widely used in colleges. His introductory text was entitled *The Elements of Moral Philosophy*, 3rd ed. (Boston and Burr Ridge, IL: McGraw-Hill College, 1999). There are numerous excellent works on approaches to Christian ethics; a recent highly readable one is written by Russell B. Connors, Jr. and Patrick T. McCormick, *Character, Choices and Community: The Three Faces of Christian Ethics* (New York and Mahwah, NJ: Paulist Press, 1998).

30. Plato's *Republic*, in which he lays out the ideal human society, is one of the first serious utopian works. Aristotle's *Nicomachean Ethics* offers a systematic approach to the ethical life. It is interesting to note that Aristotle was one of the first biologists, making detailed observations of nature, particularly animals, in works that are still referenced. Thomas Aquinas' encyclopedic *Summa Theologica* provides a systematic Christian theology infused with Aristotle's ethics.

31. Probably the most well-known Christian ethicist working within virtue theory is Stanley Hauerwas. He is a prolific writer, and one of his best volumes laying out a Christian virtue ethic is *A Community of Character: Toward a Constructive Christian Ethic* (Notre Dame, IN, and London: University of Notre Dame Press, 1981).

32. Josef Pieper's *The Four Cardinal Virtues* (Notre Dame, IN: University of Notre Dame Press, 1965) has become a classic on this subject.

33. His principal volume addressing his moral theory is entitled *Grounding for the Metaphysics of Morals*, first published in 1785.

34. Joseph Fletcher, *Situation Ethics: The New Morality* (Philadelphia: Westminster Press, 1966).

35. The principles of healthcare were developed fully by Tom L. Beau-champ and James F. Childress in their popular text, *Principles of Biomedical Ethics*, 5th ed. (New York and Oxford: Oxford University Press, 2008). For another thorough volume on these principles, see Raanan Gillon, ed., *Principles of Health Care Ethics* (Chichester, UK, and New York: John Wiley & Sons, 1994).

PART II

Genetic Technologies:
Opportunities and Challenges

4

Designer Babies: Creating Our Children

Sara and her husband Todd listened intently as the genetic counselor described a procedure to genetically modify their embryos using a technique the counselor called "preimplantation genetic modification." Todd was excited about the prospects for this new technology. This was an opportunity to have taller children who would not be ridiculed as he was earlier in life. Todd experienced the discrimination that many people short in stature encounter. He tried growth hormone therapy, but it was too late, so Todd resorted to painful surgery to extend his leg bones. Todd is nearly normal height now, but this genetic counselor offered a new and different approach for his children. The plan was to begin with in vitro fertilization and use an experimental laboratory procedure that their counselor called homologous recombination. So-called tallness alleles would be inserted into several of their early embryonic cells, each still capable of producing a healthy embryo for the next step, implantation in Sara's uterus. The plan sounded great from Todd's perspective, but Sara, who earned high grades in biology, was not so sure. The room got quiet. Startled, the counselor fell off the chair as Sara jumped up and shouted at her husband: "Todd, this is a stupid idea! You are willing to fool around with the DNA of our children, and THEIR children, so you can stop feeling guilty about your SHORT genes! Is this because you did not meet the height requirement for that promotion at work? Or is this because you expect several embryos to implant and you want an entire basketball team? And, I bet you want to select for all males!"

INTRODUCTION

The possibility of being able to modify our children before birth is one that many people view with happy anticipation as well as with a sense of concern and perhaps even dread. The new and future genetic technologies offer hope for both eliminating certain diseases and improving traits that already exist. Thus, Todd is excited that his children may be born without something that has adversely affected his own life, but the deeper question is whether or not he should do this. It is well beyond the scope of what is possible now: we cannot genetically modify an embryo so that it will be taller, especially with regard to a characteristic such as height, which we know to result from an interplay of genes and environment (such that diet, for example, can significantly affect stature). But it is already on the horizon that we will be able to modify the genes of our offspring, and possibly even future generations, and this raises some serious ethical and religious as well as biological concerns.

Many terms are used when talking about these new technologies. One of the older ones is "test tube babies," arising with the development of in vitro fertilization. This terminology arose because the union of the egg and sperm took place outside of the parents' bodies, in a Petri dish in the laboratory, with the resulting fertilized zygote ultimately implanted into the woman. Originally this technology and resulting terminology raised the specter of Aldous Huxley's *Brave New World*. His Fertilizing Room of the Central London Hatchery and Conditioning Centre contained decanters where embryos were grown, and depending on what their role in society was predetermined to be (e.g., the superior Alpha class or the lower Epsilon class), certain chemicals were added at particular points of development.

Of course we are far from this point, thankfully, but the arrival of the first test tube baby, Louise Brown in 1978 in England, marked a leap forward in technology and the subsequent increase of this now very common practice of in vitro fertilization. We also hear the term "designer children," reflective more of new technological possibilities, inherent in which is the idea that children can ultimately be fashioned somehow by their parents, given the technology available. Thus, when we talk about this kind of intervention at the genetic level, we are really talking about what is usually referred to as genetic engineering.

Implicit in the idea of genetic engineering is the recognition that we are doing something beyond simply giving birth to children; instead of "growing" them, we are "making" them.[1] Genetic engineering is moving beyond the natural birth process by which virtually all humans have been born to allow increased genetic intervention, involving a host of players other than simply the parents, such as genetic specialists and genetic counselors. Genetic engineering means that parents, at least those who have the financial means to do so, can have greater control over their children's genetic destiny and will be able to "screen in" for traits or "screen out" for disabilities in their children. It is interesting to note that scientists and ethicists tend to talk differently about the technologies and their subsequent implications with regard to embryos. Science focuses on the distinction between selection and modification. Selection has to do with screening and diagnosing embryos with regard to genetic characteristics (e.g., disease, sex), whereas modification has to do with changing or altering characteristics already present. An important distinction is also made in modification with regard to somatic therapy (which just impacts the affected individual) versus germ-line intervention, which has the possibility of altering future generations.[2] Ethics and religion, on the other hand, focus on the distinction between therapeutic and enhancement interventions. Therapeutic intervention refers to fixing what might be determined to be genetic anomalies, whereas enhancement refers to improving normal characteristics.

Taken together, science and religion/ethics must address four broad areas with regard to embryos: sex selection, screening out disabilities, enhancement, and screening in disabilities. Genetic engineering is very important because it allows us to intervene in human nature in ways not previously considered possible, except within the realm of science fiction. But as always, with new advances in science and resulting innovative technologies come additional ethical considerations. One of the most important Christian ethicists of the 20th century, Paul Ramsey, noted: "It would perhaps be better not to raise the ethical issues, than not to raise them in earnest."[3] One important concern has to do with the essence of human nature. What does it mean to be human? Is human nature fixed or malleable? At what point are we moving beyond merely the human to the "transhuman" or the "posthuman"?[4] A second consideration has to do with the boundary between the human and the divine. At what point does human action impede on what should only be left within the province of the divine? What is God's role and what is humanity's role? A third consideration has to do with how far technology should proceed. Are there some technologies that are inherently evil, or are the applications the problem? Should technologies be widely used in cases where we have not had the ability to ferret out all the ethical and religious implications? Is there a line that should not be crossed? And if so, who will decide and how will it be implemented?

To address some of these questions as well as explore the scientific possibilities that already exist, this chapter proceeds as follows. The current state of reproductive technologies is reviewed, first by providing the scientific background necessary to understand the ethical issues and then by addressing in particular two ethical issues: the role of technology, and in vitro fertilization. Second, the science of genetic engineering with regard to screening is presented, as well its resultant ethical concerns such as eugenics, playing God, and sex selection. Third, the science of genetic engineering with regard to modification is presented, as well as their resultant ethical concerns of the subsequent distinction between enhancement and therapy, choosing against a disability, enhancement, and choosing for a disability. Finally, we will offer some concluding comments.

REPRODUCTIVE TECHNOLOGIES

A wide range of reproductive options are available to individuals in developed countries; here we look at the advanced reproductive technologies that increase fertility. Fertilization occurs when a sperm penetrates an egg (called an oocyte) to form a zygote with the full genetic complement and all of the necessary materials for development of the embryo. Fertilization by natural means or in vivo (Latin for "within the living") is most common and it is generally agreed to be the most fun! In vitro fertilization (IVF) is a reproductive technology that, as the Latin phrase implies, occurs within laboratory glass. IVF is a tool for those experiencing low fertility. This basic technology begins with collecting sperm and eggs for fertilization in containers with materials to simulate natural conditions and subsequent formation of preimplantation embryos. The collection of sperm is a simple manipulative procedure that results in ejaculation and subsequent collection and preservation of the sperm. If sperm numbers are very low and it is not possible to concentrate the sperm, then it may be necessary to inject sperm directly into an egg cell (a procedure called intracytoplasmic sperm injection, or ICSI). Egg collection is more complicated than sperm collection. This is because sperm are produced regularly and are easy to access, but eggs are released on a monthly cycle and are difficult to collect. Egg collection requires hormone treatment to induce the release of eggs that are then harvested using a guided ultrasound collection device. It is also possible to harvest sperm and eggs directly from the repro-ductive tissues by biopsy, but this approach is usually limited to extenuating circumstances such as when there is a desire to obtain sperm or eggs from an incapacitated individual. In all of these cases, the intent is to increase the chances of a sperm meeting an egg while under the supervision of a medical professional. But once gametes (sperm or eggs) have been collected, gamete storage also becomes an option.

Well-developed techniques already exist for storing sperm and embryos and, recently, eggs. Advances in the low temperature pres-ervation of tissues (cryopreservation) have permitted the safe, long-term storage of sperm and the embryos created by IVF. A primary challenge was to reduce ice crystal damage to the tissues. Freezing eggs has proved more challenging than freezing sperm. Because eggs contain more water than sperm and are not fully mature until fertil-ization, they are more vulnerable in storage than sperm. A woman is born with all of her immature eggs stored within her ovaries. An immature egg begins to mature when released as part of a woman's natural reproductive cycle, and maturation of that egg is completed after penetration by a sperm. A woman's store of eggs becomes less viable as she ages and thus increases the chances of a child with Down syndrome or other birth defects. "Mothers 35 Plus" is a lead-ing Web site for late motherhood in the United Kingdom; it reports that the chances of a Down syndrome baby for a woman under 30 years of age is 1:1,529, whereas the chances are 1:112 for a woman at age 40.[5] For this reason, women do not experience the same free-dom from the so-called biological clock as do men, nor can a woman easily store her gametes when faced with chemotherapy or workplace risks with a potential to damage her internal store of eggs. That was until recent developments and companies such as extendfertility.com reported successful cold storage and subsequent fertilization of eggs.[6] Professor Ronald M. Green is the Eunice and Julian Cohen Professor for the Study of Ethics and Human Values, and Director of the Ethics Institute at Dartmouth College. Green provides a compelling argument that women's access to the freezing of their eggs, particularly young women, may be limited because of the high cost and challenges associated with the egg harvesting. To store eggs, a woman must undergo hormone treatment to release her eggs for harvesting. This is a costly and uncomfortable procedure that women might avoid except for the most compelling of reasons such as when faced with the possibility of egg damage during chemotherapy. Green suggests that developing in vitro maturation (IVM) technologies might mitigate some of these challenges. IVM begins with a relatively simple biopsy of ovarian tissue to obtain a section of ovary containing immature eggs, this tissue is then frozen using cryopreservation techniques, and the eggs may then be induced to mature as needed for IVF.[7]

An embryo cannot develop without implantation into the uterine wall for nurture by the mother. Eggs fertilized in vitro are normally transferred to the uterus with the hope that implantation and normal development will occur. The success rate for fertilization to implantation under natural conditions is estimated to be 30 to 70 percent.[8] The breadth of this estimate may be explained by the fact that it is difficult to reliably determine under natural conditions how many fertilized eggs implant and result in birth. The success rate for IVF is much easier to establish because measurable data are available at each step of the process. For example, one report from a large sample of fertility clinics indicates that when 73,406 mothers-to-be were treated with hormones, the result was 62,881 retrievals for IVF, with 59,004 subsequent transfers, of which 22,567 resulted in pregnancy, followed by 18,793 deliveries.[9] This is a 26 percent success rate and is slightly lower than the estimated 30 to 70 percent natural implantation rate. Zygote intrafallopian transfer (ZIFT) is a technique that increases the odds of implantation. This technique begins with IVF eggs, which are then transferred directly into the fallopian tubes instead of the uterus. Sperm normally swim up into the fallopian tubes to meet an egg, and it is hoped that inserting the zygote directly into the fallopian tubes will provide a more natural experience for the zygote, and thus increase chances for implantation when the zygote reaches the uterus. Gamete intrafallopian transfer (GIFT) goes a step further in mimicking nature when sperm and eggs (instead of the zygote as in ZIFT) are inserted directly into the fallopian tubes with the hope that natural conditions will result in an even greater chance for fertilization and ultimate implantation. GIFT and ZIFT procedures are less than 1 percent of all IVF procedures, and their success rates are only 22 percent.[10] Success rates may not be the best indicators for these technologies. It is likely that these parents cannot have children by natural means, and for that reason intrinsic values must be factored into our decisions when considering these technologies.

Two ethical issues about the use and expansion of reproductive technologies must be addressed: Are there limits to the use of technology in general, and are there ethical problems with IVF? With regard to technology, the key question is a boundary one: How far should we go? We can probably all agree with Philip Hefner, who has written widely on religion and science issues, that "our culture is irretrievably technological."[11] This is true in virtually all contemporary areas of human endeavors, but certainly when it comes to scientific advances dealing with genetics. It raises the broader question of what our relationship should be with technology.

Ian G. Barbour defines technology as "the application of organized knowledge to practical tasks by ordered systems of people and machines,"[12] and subsequently in the same volume offers a very helpful threefold paradigm for possible views of technology.[13] The first is "technology as liberator." This is an optimistic view that tends to welcome technological innovations as having the potential for freeing us from negative realities such as hunger, disease, and poverty. Those defending this view point to the benefits of higher standards of living, increased opportunity for choice, more leisure time, and improved means of communication. Those criticizing this view point to the environmental costs, the resultant alienation from nature, and its accessibility largely confined to those with significant financial means. The second is "technology as threat." This is a more pessimistic view held by those who consider technology to be a threat to authentic human existence. They point to the problems of uniformity and impersonality in society, a narrow view of efficiency, alienation of the worker, and the development of dangerous military weaponry. There have been several famous critics of technology, one of the foremost being Jacques Ellul. In his classic volume, *The Technological Society*, he maintains that the boundaries between science and technology have been blurred, and raises concerns about our often uncritical embrace of technology.[14] Challengers to these pessimists say that we must make important distinctions among technologies and not simply lump them all together, as well as realize that technology can be the servant rather than the architect of human values. The third view, held by Barbour himself, is "technology as an instrument of power." This view portrays technology as a neutral power, whose designs can be determined only by its social context, such that technology itself is socially constructed and can thus be redirected. These contextualists, in contrast to the optimists and pessimists, tend to focus on social justice, especially in terms of who benefits financially from the technologies developed, as well as on seeking greater environmental protection.

The view that we hold with regard to technology largely affects our attitude toward the new developments, particularly genetics, as a result of scientific advances that will continue to face us. If we are optimists, we might too quickly embrace technologies without critically assessing their potential harm to human beings. If we are pessimists, we might be too quick to reject technologies that may contribute significantly to human living and thriving. If we are contextualists, then we will proceed cautiously, realizing that the use to which we put technologies is more problematic than the actual technologies themselves.

The next issue has to do with the morality of in vitro fertilization. Fertilization takes place outside of the natural means of conception, such that the sperm and egg are united outside of the bodies and then implanted into a woman. Most individuals, including Christians, do not have a problem with this technology, especially if it is used as a means to help heterosexual couples deal with infertility problems, and even for other individuals (e.g., single women wanting a child, or even gay couples). Thus, for some there is no problem with the technology, especially since it is designed to enable those wanting children (traditionally considered a "blessing") to have them. Other individuals might want to restrict the procedure to only heterosexual couples because of their view of family or marriage as being legitimate only within the confines of heterosexual coupling.

But some have raised ethical concerns about this technology. One concern is leftover embryos. If there is the possibility that extra embryos might be destroyed or experimented on, this is morally problematic for many. This gets to the deeper question of the moral status of the embryo, which we discuss more fully in the next chapter. Another concern is the slippery slope argument—that if we engage in this kind of technology, which we might even consider acceptable, what other technologies might it lead to that we would not be as comfortable with? One could argue that perhaps we are already skating down the slippery slope, since IVF is such a taken-for-granted procedure that it may indeed be paving the way for other technologies that allow us to screen embryos for sex, for example. A third concern has to do with the practice of IVF with regard to the proper role and function of sexuality within marriage. Thus, it is worth briefly exploring the Catholic Church's objection to IVF, with which some even outside of the tradition might agree.

The Catholic Church's longstanding teaching with regard to marriage and sexuality is that sexuality should be expressed only within the confines of the marriage covenant, which by definition can be expressed only between heterosexuals. Thus, sex outside of this context is wrong, such as premarital sex or homosexual sex, for example. In addition, sexuality within marriage has two purposes: unitive and procreative. The unitive aspect has to do with the idea that couples express themselves to each other in a kind of self-giving love, and the procreative aspect has to do with the idea that every sexual act should be open to the possibility of children. Any sexual activity, and medical procedure, that violates one of these purposes is considered illicit.[15] In particular, the practice of IVF, while done in the spirit of the procreative sense (e.g., that the couple wants to have children), is considered problematic because it violates the unitive side. Even if those undergoing IVF are married to each other, the Church still believes that it is problematic. In addition, one of the ways that sperm is obtained is through the act of masturbation on the part of the male, and this practice is also considered illicit by the Church because it also violates the unitive aspect of sexuality.[16] The Church is, however, supportive of any medical procedures that would assist the couple with having children as long as it does not violate this dual purpose of marriage. Given the existence of these technologies, though, the Church emphasizes that any children born as a result of these kinds of technological interventions should be welcomed as any other child, as being a child of God with all of the dignity, respect, and care that goes along with that. Thus, children should never be treated as laboratory products, even if their existence is as a result of laboratory procedures. Finally, the Church does not think that people have a "right" to have a child. Although certain medical interventions are permissible as an aid to reproductive success, it just may not be part of God's plan for a particular couple to have children.[17]

GENETIC ENGINEERING: DIAGNOSIS AND SCREENING

In this section we explore primarily techniques that enable us to select for or against certain embryos, and in the next section we focus on modifying embryos as a result of diagnosis and screening. With both selection and modification, we are at a place in medical genetics where we might know in advance what kinds of medical problems an individual actually has or the dispositions they may have to certain diseases, and after this, it seems logical that we would try to "fix" these problems. With regard to diagnosis of and screening for disease, in particular, we have made considerable progress. Genetic screening is becoming a more widespread practice in medicine in general and not just with regard to embryos. We have increased knowledge about the actual markers on genes and chromosomal abnormalities that identify at-risk or already affected individuals. Out of the 5,000 possible genetic disorders that humans can have, we have already identified a genetic basis for about 2,000 of them. Of course, to get one's genome tested for all these diseases is prohibitively expensive for most people. Yet some limited screening (such as for a particular disease or a small group of diseases) occurs in a number of different contexts. Embryos and fetuses are often screened through a variety of techniques to establish the existence of abnormalities with the possibility of nonselection or abortion. Newborns are tested for a limited number of diseases shortly after birth so that they can best be treated medically as soon as possible. Adults can be genetically screened to confirm a diagnosis, to determine if they are a carrier of a particular genetic tendency (usually in cases where there is a family history), or to test their predisposition to particular diseases. But whenever we are talking about genetic engineering in any form, we are treading in a place where most people feel somewhat uncomfortable. After exploring the science of these diagnostic and screening techniques, we turn to some of the ethical concerns that these practices raise; we look particularly at the concept of eugenics, the issue of whether or not we are "playing God," and sex selection.

The use of IVF technologies has increased since their inception in the late 1970s, and it is likely that techniques to screen gametes and embryos will also become more effective and commonplace. Embryo screening before insertion into the uterus is the focus of this discussion. It is possible to harvest gametes and screen for genetic traits before IVF. The sorting of sperm may have been the first reliable gamete screening technique developed in the animal husbandry industry to separate male-producing sperm containing a Y chromosome from female-producing sperm containing an X chromosome. The focus here is the embryo because it is probably more efficient to screen an embryo instead of many sperm and a few eggs. At the eight-cell embryonic stage (about 3 days after fertilization), it is possible to remove a cell from the embryo and still allow for normal development without apparent harm. That cell can then be examined for a specific genetic trait, sex, and other characteristics such as chromosome number, so a decision can be made as to whether or not to insert the embryo in question into a woman. This procedure is called preimplantation genetic diagnosis (PGD) and is frequently used to determine the presence of deleterious alleles. For example, Huntington chorea is a form of dementia caused by an autosomal dominant allele in late adulthood.[18] If one member of a couple tests positive for the presence of this allele, then the chances of their having a child that will develop Huntington disease later in life is 50 percent. PGD would enable the couple to select one of their embryos that did not carry the Huntington disease allele. This technique assumes that the cell selected is representative of the remaining cells in the embryo, and it is most reliable for genetic diseases like Huntington chorea, which has a high degree of genetic heritability, but this technique is not as reliable for traits influenced by the environment such as obsessive-compulsive disorder.[19] PGD targets specific genes for embryo selection, but preimplantation genetic screening (PGS) goes beyond PGD and scans the genetic material of a cell that has been extracted from a preimplantation embryo for a range of genetic traits. PGS is readily available and enables a couple to have their embryos scanned for a range of genetic diseases (cystic fibrosis, muscular dystrophy, Tay-Sachs, and Down syndrome, to name a few).[20]

It remains to be seen to what extent couples will opt to select for traits that are not relevant to the health of their child. Two factors may contribute to an increased use of PGS. First is a rise in voluntary personal DNA screening by adults, and second are recently improved privacy and nondiscrimination laws. It has become very easy to obtain your genetic predisposition for many diseases and traits through personal genetic screening by fee-for-services companies. One such company is deCODEme.com; founded in 1996, deCode has been developing and analyzing a genealogical DNA database that includes a majority of the adult population in Iceland and thousands of people around the world.[21] Another example is Navigenics, a company that will scan your DNA for more than 2,000 genetic markers correlated with various health conditions and then provide access to a genetic counselor to interpret the results.[22] To collect DNA, these companies mail the client a kit containing the necessary directions and materials to swab and collect cheek cells for return to the company and subsequent analysis. These companies then scan the client's DNA sample for genetic predispositions to Alzheimer disease, breast cancer, colon cancer, glaucoma, macular degeneration, prostate cancer, multiple sclerosis, type II diabetes, and glaucoma, to name a few. The typical cost is $1,000 to $2,000, depending on the range of services, but as competition increases, the cost is likely to come down. This easy access by some adults may result in a greater desire by couples to select PGD and PGS. These couples have already demonstrated a desire to know their own genetic predispositions, and it makes sense that they will also desire to know the genetic predispositions for their embryos. This raises concerns about the market-driven nature of these technologies.

A second factor that may contribute to a rise in PGD and PGS is a perception that adequate privacy and nondiscrimination laws are in place. Companies that screen DNA promote their secure measures and privacy.[23] President George W. Bush signed the Genetic Information Nondiscrimination Act (GINA) into law on May 21, 2008.[24] This act is designed to prevent genetic discrimination by employers and insurance companies. It remains to be seen how effective this act will be, but at the very least, it may reduce any privacy and discrimination concerns of those of us willing to have our DNA scanned by companies like Navigenics. But now prospective parents may be even more willing to supplement their IVF procedure with PGD or PGS because of reduced concerns regarding their future child's privacy and potential discrimination. PGD and PGS of embryos can provide genetic and biochemical information to inform parental decisions for selecting specific embryos. Amniocentesis and chorionic villus sampling (CVS) are techniques that can provide the same information as PGD and PGS, but a significant difference is the fact that amniocentesis and CVS provide information *after* implantation of an embryo. Both amniocentesis and CVS are frequently used to diagnose birth defects, but CVS can provide the information somewhat earlier in a pregnancy and therefore provides parents with additional time to weigh their options. All of these procedures have a high degree of accuracy for identifying biochemical and genetic factors, but in all cases they place parents in the position of having to make a decision regarding the future of an embryo or fetus. Prospective parents may avoid such decisions by natural family planning (NFP). This low-tech approach does not accurately predict biochemical and genetic disorders, but it can assist a couple in their family planning by monitoring a woman's ovulation cycle so that intercourse can be timed accordingly for family planning. NFP techniques include ways to avoid pregnancy, enhance fertility, and increase the odds for a male or female child. Low-tech methods (such as the Shettles method) rely on characteristic differences in male- and female-producing sperm as a way to increase the chances of a boy or girl according to the timing and positioning during intercourse. These low-tech methods are not as reliable as high-tech techniques such as sperm sorting.[25] The options are many, but so are the ethical considerations.

One of the most common reactions to genetic engineering revolves around the concept of eugenics. Though this term literally means "well-born," it often conjures up an image of Hitler and the Nazi party in their attempts to create a master race through eliminating "undesirables." Thus, the concept of eugenics usually has a negative connotation. But in essence it means trying to improve the genetic make-up of a population through human intervention, whether genetic or otherwise.[26] A few important distinctions should be made. First, negative and positive eugenics are not necessarily the same thing. Negative eugenics has to do with eliminating undesirable conditions in humans, usually diseases, which often result in some suffering for not only the afflicted person but for their loved ones and society as well. Positive eugenics, on the other hand, is more akin to genetic modification in that the attempt is to improve on desirable characteristics already available. Thus, negative eugenics can include trying to diagnose and screen out muscular dystrophy, for example, whereas positive eugenics might include an attempt to genetically improve the intelligence of one's child. A second distinction is between what can be termed the old eugenics and the new eugenics. The old eugenics would include techniques such as forced sterilization of certain individuals whom one does not think should be able to reproduce, or eliminating individuals themselves who have certain illnesses. The new eugenics includes interventions related to some of the new technological developments such as prenatal diagnosis, germ-line therapy, selecting certain embryos with particular characteristics rather than others, or cloning.[27]

American society engaged in forms of the old eugenics in the 1920s, which included laws requiring forced sterilization of certain populations, particularly those who were mentally challenged. The Nazis also engaged in this old eugenics through their plan to eliminate those who were not physically or mentally normal or superior, at least by their standards. Some of the Nazi propaganda films in particular tried to persuade the German citizenry through vivid images why society would be better off without certain types of individuals. It is well known that in addition to Jews, the Nazis targeted gypsies, homosexuals, and the mentally and physically challenged. One of the major problems with the old eugenics, in addition to the means of discriminating between the less worthy and the more worthy, and the methods by which they were discriminated against, was the issue of coercion. Individuals in virtually all cases of the old eugenics did not have much choice about their fate. This is somewhat different from the new eugenics in which at least parents are often involved in the decision as to whether or not to select against or select for certain traits. But the deeper issues still remain as to the effect on the individuals born who had no choice in terms of how they would be designed, and the possible hubris of parents who feel comfortable designing their children.[28]

The expression "playing God" is often brought up when genetic engineering is a topic of discussion. This phrase raises the boundary issue of where human intervention is legitimate, and when what we are doing may actually fall more properly within the province, and even providence, of God. Paul Ramsey noted: "Men ought not to play God before they learn to be men, and after they have learned to be men they will not play God."[29] But we cannot either agree or disagree with Ramsey about what humans should do if we do not agree upon what we mean by playing God. One way to approach the definition is to say what we do *not* mean by it. We do not think that humans are playing God when doctors perform medical procedures such as surgery, or when they prescribe certain antibiotics or other medications, or when a dentist pulls a sick tooth. We do not lament that as medical professionals they are overstepping their bounds. Instead, we might regard someone who made such an argument as being unreasonable. Certainly we do believe that humans can and should medically intervene in any number of medical situations. In fact, one could argue that if humans are created in the image of God, part of which means that we have been endowed with reason and intelligence, then when doctors or other medical professionals develop cures and help sick individuals, they are doing God's work by helping the sick through medical interventions developed by humans. Usually, though, when the phrase playing God is used, it often is in reference to new genetic technologies that permit human intervention in a way not previously possible or even considered within the realm of human action.

Ted Peters offers a very interesting discussion of three possible ways that the concept of playing God can be used, which he believes can have overlapping meanings.[30] The first possible and generally noncontroversial meaning deals with "learning God's awesome secrets." This has to do with the sense of awe we have when contemplating the mystery revealed by new discoveries and even new technologies. The second has to do with the wielding of "power over life and death." This encompasses the idea that part of the role of medicine is to save life, which seems logical to most of us. The third use has to do with the application of science to "alter life and influence human evolution." We are now talking about DNA and whether or not there is something sacred to DNA that should not be so vulnerable to human intervention. Thus, it is this third meaning that is often meant when the concern is raised about humans playing God. It seems that we are truly on the precipice, and are probably even halfway across it, of truly new breakthroughs and possibilities with regard to the essence of what it means to be human. Genetic engineering does indeed raise the question of whether or not we are doing something different now with our new genetic technologies, and whether or not we should truly tread with caution, if at all, into God's territory. Another way of trying to distinguish between the human and divine role in medicine has to do with whether God has given humans the role as stewards or cocreators in terms of our responsibility to the rest of the created order. Traditionally the first few chapters of Genesis have been interpreted as God having given Adam and Eve the responsibility of being stewards, which means being responsible for what God has already created. But some contemporary theologians think that God has given humans the responsibility of not only caring for something that already exists, but also being directly involved with God in helping to create new realities and possibilities. Thus, humans can be considered cocreators with God, including within the realm of medicine.[31] In any case, when using the phrase playing God, it is important to be clear about what we mean as well as recognize that virtually all new medical technologies have the potential to raise the concern about whether or not we have crossed this line. But certainly we should probably proceed a bit more cautiously now that we have the increasing potential to affect not only individuals but also future generations with our genetic interventions.

A more specific concern related to genetic diagnosis and screening is sex selection. We now have procedures that enable us to virtually ensure the implantation of an X versus a Y embryo. The increased chances of having a boy or a girl are also attributable to more natural methods, such as the timing of intercourse to ensure at least a greater probability of either a boy or a girl. In addition, abortion has been used in some cultures upon discovery that the fetus is a particular sex and the parents do not want a child of that sex. Of course, this practice raises the same kinds of concerns that are raised by technological selection procedures—should we be selecting for sex at all? There are several possible reasons why parents may choose one sex over another. One reason is family balancing: this is the attempt to balance out the number of boys and girls one has, regardless of the number of children one has. A second reason is that parents may want to have a particular birth order, with either a boy or girl coming first. A third reason is that some cultures favor boys over girls for economic or other reasons, and thus girls may be more likely to be selected against.

Obviously, some individuals have no problem with having the option of sex selection, and some will certainly make this kind of selection. In favor of this are approaches in liberal eugenics, which maintain that genetic decisions about offspring properly belong to parents. If parents want to choose for a boy over a girl, or vice versa, then they should be free to do so. There are, however, some concerns raised with this practice. Nature left to its own devices usually winds up with approximately a 50–50 ratio with regard to sex, and human intervention may throw off the balance. In cultures where this sex selection is more routinely practiced than in ours (such as in India and China), the imbalance in numbers has significant implications when these individuals are old enough to marry. Thus, there are not enough women for all of the men who want wives. An additional concern is that sex selection tends to work against women, so that boys are often selected over girls, especially in cultures that have strong economic incentives to do so. However, Michael J. Sandel raises the interesting question of whether or not we might view sex selection differently if it did not affect the sex ratio.[32] Another question is whether or not we should be making these kinds of selections. After all, parents for much of human history did not usually care if their child was a boy or a girl, as long as it was healthy. Ultimately, though, individual choices accumulate and can have social implications. Whatever we believe about sex selection and the technologies that make it easier to happen, we cannot simply approach this or any other genetic technological possibility only in terms of individual choice; we must consider the larger social impact.

GENETIC ENGINEERING: MODIFICATION

Will couples also be willing to manipulate the genetic information in their embryos? The technology for genetic modification of embryos builds on methods developed for IVF, PGD, and PGS. Thus, in the future, preimplantation genetic modification (PGM) will become an attractive option for some couples. But first we need to pay more attention to the human subject research phase of PGM. Rebecca Susan Dresser is the Daniel Noyes Kirby Professor of Law and Professor of Ethics in Medicine at Washington University, and she makes several policy recommendations for oversight of PGM research.[33] Dresser suggests that the current push for PGM research may be a result of limited success with human somatic cell interventions as therapies for genetic diseases. She emphasizes the need for policies that protect the "later born children" of genetically modified embryos (the next generation). Publicly funded research for these technologies is reviewed by the Recombinant DNA Advisory Committee (RAC) of the NIH.[34] And recently the Food and Drug Administration (FDA) also claimed "jurisdiction over human cells used in therapy involving the transfer of genetic material by means other than the union of gamete nuclei."[35] Dresser calls for RAC and FDA efforts to consider PGM and future generations. She argues that the current practice of viewing a child's welfare as the primary objective of PGM research does not adequately protect "later born children." A great deal of federal policy and ethical debate has focused on the destruction of embryos; and now the research has moved beyond IVF, PGD, and PGS to include genetic modification of embryos with the potential to alter future generations.[36] Couples growing up in an era of voluntary genetic screening might also desire the same for their embryos, but is PGM research prepared to support real-world applications? The answer is yes for animals, but not for humans. Hundreds of transgenic animals have already been created (meaning that they contain genes foreign to their genome), and many of those are patented. For example, Harvard University's "OncoMouse" contains human cancer genes, and it may serve as a model organism in research.[37] Transferring genes to another organism can be an expensive and complex task, but the need for human gene therapies and other practical applications has driven the technology. Transgenic animals can produce rare gene products and human organs with fewer rejection concerns, yet the regulatory processes and ethical considerations remain unresolved (this is even a bigger hurdle for the genetic modification of human embryos). To date, the FDA has not approved any gene therapy techniques,[38] and it is taking steps to seriously consider the risks and benefits of xenotransplantation (replacement of human organs with animal organs).[39] According to some researchers, there is a need for well-established embryonic stem-cell lines in animals such as pigs and cows to facilitate research and development of technologies for the genetic modification of embryos.[40]

A range of techniques exist for inserting genes into cells and ultimately embryos, but the techniques differ in their precision. We consider these techniques in three broad categories: direct insertion of DNA into cells; the use of vectors to carry DNA into cells; and homologous recombination that may improve precision once DNA has been inserted into cells. First, it is possible to manually introduce genes into cells by microinjection, liposome transfer, and other ways, with the hope that the genes will take up residence at functional locations in the cell's genome. Early on, these methods were sometimes called "shotgun approaches" because they lacked precision, but recent technologies may change this perception. Direct microinjection of DNA into zygotes has been used to create transgenic animals, but it is not yet practical in humans because it is too inefficient. The

DNA integrates randomly in the animal's genome and thus mutations can result, and there is no guarantee that all cells will be modified. Liposome-mediated transfer is when membrane-bound lipid (fat) vesicles carry drugs or genes into cells. Large doses can be delivered, but limitations include short shelf-life, lack of tissue specificity, and rapid clearance from the body. Examples of current use include the treatment of HIV-associated Kaposi sarcoma and the delivery of antibodies against oncogenes in breast cancer.[41]

A second approach includes vectors that carry and deliver genes to cells. Vectors include artificial chromosomes and viruses.[42] A vector behaves like the legendary Trojan horse, but instead of carrying soldiers into the city of Troy, a vector carries DNA into an organism's cell. Artificial chromosomes can act as "imposters" when they are inserted into cells and then are mistaken for normal chromosomes in the cell with subsequent replication and gene expression. Yeast artificial chromosomes (YACs) have been manipulated since 1987, and thus the technology to build artificial chromosomes is well developed, but human artificial chromosomes (HACs) are much larger and therefore difficult to manipulate. Efforts are under way to develop mini-chromosomes by either reducing the size of an existing human chromosome or building up from a smaller chromosome.[43] Adding an intact chromosome pair to the cell might trick it into managing the chromosome and its genes just like one of its own for cell replication and the expression of traits. This technique may someday serve to add an array of genes that could add entirely new processes to cells. Virus vectors are also a method for delivery of genes to human embryonic stem cells.[44] Viruses have coevolved with cells, and viruses thus already have mechanisms to insert their genetic material into a specific cell type for manufacture of more viruses by the cell's biochemical processes. The advantage of using viruses is that they infect specific tissues and can be modified so as not to cause disease while delivering the DNA, but this process is not always as precise as it sounds, and unintended consequences may occur.

The third technique, homologous recombination, may make it possible to repair a specific gene in cells at a specific location in the cell's genome. The gene can then be replicated, transcribed, and translated by natural processes. Homologous recombination techniques show promise as a way to insert genes into cells with the precision necessary for normal function in subsequent cells. Recombination of homologous chromosomes is a naturally occurring event in germ cell production that shuffles the genetic material between chromosome pairs. A section of DNA from one chromosome arm is exchanged with a corresponding section of DNA on the other arm. Scientists are seeking ways to exploit this natural process as a way to trick cells into substituting a DNA sequence of interest into its existing genome. Homologous recombination techniques build on nature's repair mechanism to reconnect the exchanged sections of DNA. Laboratory studies have demonstrated specificity in that foreign DNA can replace single-base mutations at a 1–2 percent correction rate in human cells.[45] Homologous recombination is a long way from providing Todd and Sara the opportunity to introduce "tallness alleles" into their embryos. Given the science today, however, it is not too far-fetched to imagine the following scenario. The medical team would begin by sampling and analyzing Todd's DNA so they could then synthesize small sections of DNA that are compatible to regions adjacent to Todd's so-called "short genes" (Todd might have elected to modify his sperm, but in this case, Sara and he were already familiar with the IVF procedure). The medical team could then insert the synthesized sections of DNA directly into their embryonic cells. In addition to tallness alleles, a small section of DNA would be synthesized to act as a probe to seek out similar sequences that were previously identified in Todd's DNA as being near the loci for stature. The probe would then integrate into the embryonic DNA and trick the cells into a repair process that would add tallness alleles. It is important to note that it is unlikely that all of the cells would be modified. Stem-cell research may provide ways to produce embryos with all of the cells modified, but for the case imagined here, it is assumed that not all of the cells need to be converted because stature is a quantitative trait that is also influenced by environmental factors such as diet.

If a genetic modification or gene insertion is to benefit the recipient, and also be transmitted from generation to generation (a germ-line intervention), then all cells of the embryo must be successfully targeted. Otherwise, as the embryonic cells divide, multiple cell lines will develop, and it is left to chance as to whether or not reproductive tissues will contain the modification. Embryos and resulting organisms with multiple cell genomes are called chimeras. The long-term effectiveness of techniques that remove cells from embryos (normally at the 8-celled stage) for subsequent modification and reconstruction of an embryo without apparent harm remains to be demonstrated. This technology is well developed for use in mammals and might pave the way for the use of PGM in humans. Producing chimera embryos may be adequate when dealing with recessive diseases such as cystic fibrosis where adult symptoms can be reduced if at least some of the cells express the normal version of the gene. But this may not be adequate for diseases such as Huntington where a dominant gene causes the disease, and where it is not clear that modifying only some of the cells will be adequate to alleviate symptoms. In the case of Huntington, it would probably be more efficient to use PGD and select for an embryo without the Huntington gene.[46] The reality is that germ-line modification is not yet an option for humans, but the technologies are already being applied in animals.

Gene therapy differs from conventional medical procedures in that it targets the gene or genes responsible for a particular medical condition. It makes a big difference as to whether or not the genes targeted are in somatic or germ-line cells. Gene therapies in children and adults target genes in the somatic cells of specific tissues and organs to mitigate a disease; this is somatic-cell line modification (SLM). Disorders that affect the whole body, such as cystic fibrosis and Duchenne muscular dystrophy, are logical choices for germ-line modification (GLM). Couples who are both asymptomatic carriers for a recessive disease allele such as cystic fibrosis have a one in four chance of having a child with the disease. Currently, these parents have three options: they could decide not to conceive, they could use genetic testing to screen the fetus followed by selective termination, or they could use IVF and PGD to select a healthy preimplantation embryo.[47] PGM may thus become a readily available option for humans. Then, we will be designing not only our children but their children as well, and generations to come. It is important to consider how SLM and GLM differ.

Both SLM and GLM therapies are similar in that they have two primary challenges: precision and expression. Precision requires that the gene or genes of interest be accurately placed, replaced, or modified in the targeted cells or tissues or both. Functional gene expression requires that a gene, in addition to having a correct DNA sequence, be located in the cell's genome at a place where the gene can be turned on and off when appropriate to support cellular processes. Consider the following analogy. Imagine that you and your friend Jane are using the same book on home improvement, with the book representing a chromosome and the pages representing the genes. Jane drops her book in the paint while preparing to refinish the ceilings in her apartment. She checks the page index of her book, and asks you to photocopy your page 13 on how to paint ceilings. Jane then takes your clean copy of page 13 (clearly marked), but her page numbers are smudged and she is unsure which page to replace in her book. Jane then makes an effort to insert the new page at the correct location. She follows the directions, but becomes confused when directed to first uncover the floor and prepare for sanding—why uncover the floor if painting the ceiling? Jane quickly realized her new page 13 was in the wrong place. In the same way, genes that are inserted in the wrong place within a cell might work by chance, but given the multitude of genes in cells, it takes precision to place or repair genes at the correct location within the cell. If genes are placed in the wrong location on a chromosome (as was the page in Jane's book), it becomes a matter of chance as to whether or not they are expressed correctly; cells might not be as quick as Jane to recognize the problem. Genes in the wrong place might be turned on or off when they should not be, they might not work at all, or they may remain dormant with unknown future implications. For SLM, the unknown implications are limited to the lifespan of the somatic cell line and its possible effects during the lifetime of the organism. But for GLM, the unknown implications may also continue for generations to come. We must better understand the approaches for inserting genes.

Is germ-line intervention a threat to our genome and the genome of other species? If it is not precise, possibly! It is estimated that only about 2 percent of the DNA in our genome encodes for genes. It is also estimated that most of our genome originated in other species, and that genetic sequences are capable of jumping from place to place in our genome. It is also estimated that about one in eight sperm carries a new random mutation that is the result of an insertion of a DNA sequence from another location. This observed high level of random insertion as "background" mutation raises concerns regarding current attempts at somatic cell gene therapy. Current techniques for somatic cell gene therapy lack precision and may therefore introduce pieces of DNA that could integrate into our genome at random with a potential to mutate existing genes in the cell. Methods for the long-term observation of somatic cell gene therapies should be developed to determine if unintended GLM is occurring and to consider the stability of our genome. Given concerns regarding GLM, we should move forward cautiously with somatic cell gene therapies as we develop ways to treat serious genetic conditions.[48]

GENETIC MODIFICATION: ETHICAL CONSIDERATIONS

One of the debates with regard to the ethics of genetic engineering has to do with whether we are talking about therapy or enhancement.[49] In some cases the distinction is clear, so that we can agree that trying to eliminate or fix a serious physical illness such as muscular dystrophy or Tay-Sachs syndrome, for example, is obviously therapy; and trying to have a child with increased muscle tone is enhancement. But what about cases that are not so clear, such as increasing the height of a child whose extremely short stature (such as in the case of little people) might put them at a disadvantage in our society in a number of ways? Regarding enhancement technologies, nongenetic methods already exist, such as in cosmetic surgery or drug therapy for mental illnesses, but altering genes of affected individuals, whether through somatic or germ-line interventions, certainly raises the issue to a new level. We explore three different types of genetic engineering that are often discussed in terms of their ethical issues: choosing against a disability, enhancing, and choosing for a disability.

Choosing against a disability is arguably the least controversial of the three. Most of us would agree that serious physical or mental disabilities can often have a significant negative impact on the individuals so afflicted, as well as their family members. Some diseases are the result of a single genetic marker, such as cystic fibrosis, and thus are easier to identify, fix, or even screen out for (all of these options are not yet available for most genetic diseases). Other diseases have multiple factors involved, whether multiple genes, a single gene plus environment, or multiple genes and environment. For example, the predisposition for developing breast cancer has been associated with at least two different genes. Diseases such as heart disease tend to run in families, indicating a genetic predisposition, but we also know that environmental factors such as diet and exercise can significantly affect the expression of this disease. But from an ethical and religious perspective, we need to ask several questions. First, what is the role of medicine and healing? Certainly from a Christian perspective, illness has traditionally been considered a result of the fallen state of the world, which is not part of God's ideal plan. In addition, healing was integral to the ministry of Jesus and his disciples, and healing ministries have been part of the Christian tradition and ministry throughout history, such as in the establishment of hospitals and other institutions to help the sick. Prayer for the sick is part of every Christian tradition. But we can ask whether the foray of medicine into genetic engineering truly constitutes healing or belongs to an entirely different realm. Second, what do we actually mean by the term "disability," and how does a disability differ from normal variation within the human species? For example, left-handedness is found in a minority of the population, but this does not make it a disability, even though it can produce difficulties for the left-handed individual trying to maneuver in a right-handed world. We clearly agree that Huntington is a disease, however, and part of what makes it so is the significantly diminished quality of life that a person with such a disease experiences.[50] This question gets at the essence of what is "normal" and even how important that concept is. Third, one can ask the question whether it is God's will for certain individuals to be born with disabilities. Whether or not one believes that God causes or simply permits such individuals to be born, many in the Christian tradition consider the birth of a child with a disability to be a special blessing from God that God entrusts only to those who have the resources needed to care for such a child. This does not necessarily mean that God thinks it is a good idea for people to suffer, and certainly any attempt to help those with disabilities is laudable. However, it also means that God's plan might include suffering for humans, and even if one does not want to go that far, there undoubtedly exist some limits in terms of what we can do to eliminate disabilities. Fourth, who should make the decision as to whether a particular life is or is not worth living? For Christians, this decision has usually been considered to be within the province of God, and yet new technologies and practices, even issues such as euthanasia, raise new questions about human intervention in life, death, and health. Finally, eliminating disease sends a mixed message to those already living with disabilities. One could argue that we are sending the message that although we welcome the already-born with disabilities into our community, we want to prevent future members from entering in.[51]

Enhancement raises more troubling issues than choosing against disabilities. We do have the problem of differentiating between therapy and enhancement.[52] But even if we agree on what enhancement is, we still have to decide whether it is legitimate. Obviously we do engage in enhancement in several nongenetic ways already: we perform cosmetic surgery to fix "defects" or to improve on natural characteristics; we prescribe medication that can alter an individual's personality; and parents regularly engage in efforts to improve their children's chances in life in a number of ways. The question is, Is there a difference between these methods and trying to optimize a child's opportunities through genetic technologies? We consider briefly the more common arguments for and against genetic enhancement. Those in favor of it maintain that we already do this. Many parents try to improve their child's educational and ultimately professional opportunities by sending them to special schools, by enrolling them in classes to prepare them for standardized tests such as the SATs, or by sending them to sports camps to provide them with a competitive edge. An additional argument is that we live in a free market economy and if technologies are available and parents can afford them, then why not give their child every opportunity?[53] Finally, some maintain that we do applaud those who try to better themselves naturally—those who diet and lose weight, those who engage in rigorous athletic training programs to perfect their ability, and those who diligently apply themselves to study to achieve their professional goals. Arguments against genetic engineering include the major concern that only those with the financial means to do so will be able to avail themselves of these technologies. It raises the question of justice, which has to do with the fair distribution in a society of burdens and benefits. The Christian God is ultimately understood as one who loves all equally and desires the fair treatment of all of God's children. Another argument has to do with the issue of unfair advantage. Given the fact that this technology will be available only to some, the benefit to them will create an unfair disadvantage to others. Some are concerned that our understanding of and striving for human perfection reflect social ideals and constructs with which not everyone will agree.[54] For example, although thinness is valued in our society, this has not always been the case even in our society, let alone in other societies. Concerns are also raised about what the essence of human nature is and whether or not we are moving toward a posthuman nature, especially with the possibility of germ-line interventions that would affect future generations. From a religious perspective, will we be adjusting things that have no need of adjustment?

Choosing for a disability may seem like a strange choice, and yet a recent case about a deaf lesbian couple desiring a deaf child challenges the notion of what a disability is, as well as whether parents with what has traditionally been considered a disability should be able to choose that for their children.[55] The couple in question desired a deaf child and tried to increase their chances of this happening by selecting a deaf sperm donor. They argued that deafness is not a disability, but rather a cultural identity. In addition, they wanted to have a child like themselves, just as most parents do. They argued that society's discrimination against the deaf was the problem and not deafness itself. Not only the deaf community, but also little people, the blind, and others with disabilities that generally do not affect cognitive ability challenge society's notion of what is normal. They maintain that the problem is not so much with their so-called disability, but with society's perception of them. Finally, the deaf couple thought it would be easier for them to raise a child more similar to themselves than they could with a hearing child. What makes the case even more controversial was that when their son was born, he did have some residual hearing, but the parents refused to use any technologies that would enable him to further develop his hearing capacity. Several arguments have been raised against parents being able to choose for what have traditionally been considered disabilities, which can be applied to this case. First, although the deaf may not consider deafness to be a disability, there are certainly significant disadvantages to being deaf in a largely hearing society: educationally, professionally, and socially. Second, do parents have the right to intentionally bring a child into the world that will make life harder not only for the child but also create a burden on society, which will have to provide special accommodations for this individual? Third, although the deaf should certainly not be discriminated against, there is a big difference between trying to accommodate for the special needs of those who happen to be born deaf and actually attempting to create a deaf child. Finally, this situation raises the question of how much choice any parent should have in the design of their offspring.

CONCLUSION

The issues involved in genetic engineering are complex, and yet the technologies will continue to be developed, and marketed to the public, before we have time to work carefully through all the ethical and religious minefields. Although most of us agree that trying to eliminate diseases is a good idea (although some definitely prefer to do this only if there is no destruction of embryos), the possibility of enhancing traits in children is much more problematic. In some cases, we are still in the nascent stages, and although technologies for muscle development and memory improvement are on the horizon, technologies to increase the intelligence or even the height of one's child are much more distant possibilities. In fact, we may never be able to make these kinds of changes genetically with traits that truly are multifactorial, which involve both nature and nurture, both multiple genes and the environment. But of course, even if we are able to do these things, the persistent questions remain. There are certainly those who argue for a liberal eugenics in which parents should be able to make these decisions for their children. But others continue to wonder: Should we be doing this? Is there truly a difference between growing and making our children? What values should guide our discussion? Are there indeed universal traits that we can all agree on that would be good for all to have? What effect will all of this have on the programmed children that we eventually produce? And of course, the justice question always looms in the background of who will actually have access to these technologies. Whether we will one day live in a society somewhat akin to Huxley's *Brave New World* has yet to be seen, but proceeding cautiously would probably be a good beginning!

NOTES

1. Jürgen Habermas, *The Future of Human Nature* (Cambridge: Polity Press, in association with Blackwell Publishing, 2003), 47–53.

2. For an excellent resource on religious responses to germ-line modification, see Ronald Cole-Turner, ed., *Design and Destiny: Jewish and Christian Perspectives on Human Germline Modification* (Cambridge, MA, and London: MIT Press, 2008).

3. Paul Ramsey, *Fabricated Man: The Ethics of Genetic Control* (New Haven, CT, and London: Yale University Press, 1970), 13.

4. This term "posthuman" has been developed to correspond to the reality of our postmodern world; see Brent Waters, *From Human to Posthuman: Christian Theology and Technology in a Postmodern World* (Burlington, VT: Ashgate, 2006).

5. http://www.mothers35plus.co.uk/down.htm (accessed June 2, 2008).

6. For extensive information on human egg cryopreservation, see http://www.extendfertility.com (accessed May 24, 2008).

7. Ronald M. Green, *Babies by Design: The Ethics of Genetic Choice* (New Haven, CT, and London: Yale University Press, 2007), 49–52.

8. T. G. Kennedy, "Physiology of Implantation," 10th World Congress on In Vitro Fertilization and Assisted Reproduction, Vancouver, Canada, May 24–28, 1997, http://publish.uwo.ca/~kennedyt/t108.pdf (accessed May 25, 2008).

9. R. Lewis, *Human Genetics: Concepts and Applications*, 8th ed. (New York: McGraw-Hill, 2008), 417.

10. Ibid.

11. Philip Hefner, *Technology and Human Becoming* (Minneapolis: Fortress Press, 2003), 60.

12. Ian Barbour, *Ethics in an Age of Technology*, Gifford Lectures, vol. 2 (San Francisco: HarperSanFrancisco, 1993), 3.

13. The following discussion of his paradigm can be explored more fully in his volume on pages 3–24.

14. Jacques Ellul, *The Technological Society*, trans. John Wilkinson (New York: Alfred A. Knopf, 1964).

15. Thus, the *Catholic Catechism* (1994) explains it as follows: "The spouses' union achieves the twofold end of marriage: the good of the spouses themselves and the transmission of life. These two meanings or values of marriage cannot be separated without altering the couple's spiritual life and compromising the goods of marriage and the future of the family" (par. 2363).

16. *The Charter for Health Care Workers* (Pontifical Council for Pastoral Assistance, originally published in 1995) states: "Every means and medical intervention, in the field of procreation, must always be by way of assistance and never substitution of the marriage act" (par. 22).

17. For additional discussion of the Catholic Church's views on reproductive technologies and other issues related to genetics, see John F. Morris, ed., *Medicine, Health Care, and Ethics: Catholic Voices* (Washington, DC: Catholic University Press of America, 2007); and Thomas A. Shannon and James J. Walter, *The New Genetic Medicine: Theological and Ethical Reflections* (Lanham, MD, and Boulder, CO: Rowman & Littlefield, 2003).

18. OMIM, http://www.ncbi.nlm.nih.gov/entrez/dispomim.cgi?id=143100 (accessed May 24, 2008).

19. OMIM, http://www.ncbi.nlm.nih.gov/entrez/dispomim.cgi?id=164230 (accessed May 24, 2008).

20. For example, see the West Coast Fertility Centers at http://www.ivfbaby.com/AI-New-Technology-PGD.aspx (accessed May 24, 2008).

21. http://www.decodeme.com/index/about_decode (accessed May 25, 2008).

22. http://www.navigenics.com/healthcompass/MembershipBenefits (accessed May 25, 2008).

23. See Navigenic's privacy policy at http://www.navigenics.com/policies/Privacy (accessed May 25, 2008).

24. http://www.genome.gov/24519851 (accessed May 25, 2008).

25. Some low-tech options are http://www.in-gender.com/Gender-Selection/Shettles and http://www.babyhopes.com; some high-tech options are http://www.gender-selection.com and http://www.microsort.net.

26. Paul Flaman, *Genetic Engineering, Christian Values, and Catholic Teaching* (New York and Mahwah, NJ: Paulist Press, 2002): 89.

27. Ibid., 92–94.

28. A very useful volume addressing the concerns raised by the new eugenics is written by Michael J. Sandel, *The Case against Perfection: Ethics in the Age of Genetic Engineering* (Cambridge, MA, and London: Belknap Press of Harvard University, 2007).

29. Ramsey, *Fabricated Man*, 138.

30. The following discussion is based on his book, *Playing God? Genetic Determinism and Human Freedom*, 2nd ed. (New York and London: Routledge, 2003), 11–15.

31. A helpful collection of essays that address this distinction can be found in Ronald Cole-Turner, ed., *Jewish and Christian Perspectives on Human Germline Modification* (Cambridge, MA, and London: MIT Press, 2008).

32. Sandel, *The Case against Perfection*, 3–4.

33. R. Dresser, "Genetic Modification of Pre-implantation Embryos: Toward Adequate Human Research Policies," *Milbank Quarterly*, 82, no. 1 (2004): 195–214.

34. http://www4.od.nih.gov/oba/rac/aboutrdagt.htm (accessed May 26, 2008).

35. U.S. Food and Drug Administration, Center for Biologics Evaluation and Research, letter to Sponsors/Researchers: Human Cells Used in Therapy Involving the Transfer of Genetic Material by Means Other Than the Union of Gamete Nuclei, from Kathyrn C. Zoon, Ph.D., Director, updated May 13, 2002, http://www.fda.gov/CBER/ltr/cytotrans070601.htm (accessed May 26, 2008).

36. Dresser, "Genetic Modification," 195–214.

37. http://www.wipo.int/wipo_magazine/en/2006/03/article_0006.html (accessed April 27, 2009).

38. http://www.fda.gov/cber/gene.htm (accessed June 1, 2008).

39. http://www.fda.gov/cber/xap/xap.htm (accessed June 1, 2008).

40. C. L. Keefer, L. Blomberg, and L. Talbot, "Challenges and Prospects for the Establishment of Embryonic Stem Cell Lines of Domesticated Ungulates," *Animal Reproduction Sciences* 98 (October 14, 2006): 147–68, http://www.ars.usda.gov/research/publications/publications.htm?SEQ_NO_115=188028 (accessed June 1, 2008).

41. Bhavani G. Pathak, "Scientific Methodologies to Facilitate Inheritable Genetic Modifications in Humans," in *Designing Our Descendants: The Promises and Perils of Genetic Modification*, ed. Audrey R. Chapman and Mark S. Frankel (Baltimore: Johns Hopkins University Press, 2003), 55–67.

42. Plasmids as vectors are not discussed here but should be mentioned as containing so-called "naked DNA" that is not packaged with protein as in the common chromosome. Bacterial plasmids are small circular pieces of DNA that can transfer genes directly between bacteria. For example, the transfer of plasmids between bacteria is a natural process that carries antibiotic resistant genes from one bacterium to another.

43. Pathak, "Scientific Methodologies," 55–67.

44. N. Zaninovic, J. Hao, J. Pareja, D. James, S. Rafii, and Z. Rosenwaks, "Genetic Modification of Pre-implantation Embryos and Embryonic Stem Cells (ESC) by Recombinant Lentiviral Vectors: Efficient and Stable Methods for Creating Transgenic Embryos and ESC," *Fertility and Sterility* 88, Supplement 1 (September 2007): S310.

45. Pathak, "Scientific Methodologies," 55–67.

46. Green, *Babies by Design*, 46–48.

47. R. Michael Blaese, "Germ-Line Modification in Clinical Medicine: Is There a Case for Intentional or Unintended Germ-Line Changes?" in *Designing Our Descendants: The Promises and Perils of Genetic Modification*, ed. Audrey R. Chapman and Mark S. Frankel (Baltimore: Johns Hopkins University Press, 2003), 68–76.

48. Ibid.

49. For an excellent discussion, see a series of essays on this topic in Erik Parens, ed., *Enhancing Human Traits: Ethical and Social Implications* (Washington, DC: Georgetown University Press, 1998).

50. There are many good resources for exploring this issue of a what is a disability; see Neil G. Messer, "The Human Genome Project, Health, and the 'Tyranny of Normality,'" in *Brave New World: Theology, Ethics, and the Human Genome*, ed. Celia Deane-Drummond (London and New York: T&T Clark, 2003); and C. M. Culver, "Concept of Genetic Malady," in *Morality and the New Genetics*, ed. B. Gert (Boston: Jones & Bartlett, 1996), 147–66.

51. A very good volume addressing the concerns of the disabled can be found in John Swinton and Brian Brock, eds., *Theology, Disability, and the New Genetics: Why Science Needs the Church* (London and New York: T&T Clark, 2007).

52. For a good discussion on enhancement, see John Harris, *Enhancing Evolution: The Ethical Case for Making Better People* (Princeton, NJ: Princeton University Press, 2007).

53. Two writers who advocate for this approach in their work are Nicholas Agar, *Liberal Eugenics* (Malden, MA and Oxford: Blackwell Publishing, 2004), and Green, *Babies by Design*.

54. An excellent work that critiques this notion of perfection is Sandel, *The Case against Perfection*.

55. Liza Mundy, "A World of Their Own," *Washington Post*, magazine section, March 22, 2002.

5

Searching for the Fountain of Youth: Stem-Cell Research

Clint and Marian were lifelong partners, no longer interested in having children. They were now asking for the return of Marian's ovarian tissue that was in cold storage at a fertility clinic. The clinic counselor was curious and asked why she wanted her ovarian tissue after so many years. "We want eggs for life!" she said, and then proceeded to describe their plan. Clint had heard of an experimental procedure called somatic cell nuclear transfer. The procedure would use Mary's eggs to reprogram adult cell nuclei for the production of stem cells that are capable of becoming all types of cells, tissues, and someday even organs, all with little chance of rejection for the adult stem-cell donor. Marian and Clint could easily afford to pay for the development of their own personalized stem-cell lines for regenerative therapies that could extend and improve the quality of their lives. Shocked by what she heard, the genetic counselor blurted out, "We do not store eggs for the purpose of developing back-up tissues!" Marian was outraged when the counselor refused to release her ovarian tissue. "These are my eggs you are storing; nowhere in our agreement did your clinic indicate that I could not use the eggs for a purpose other than reproductive assistance! Why should we be forced to use an egg donor when my eggs are readily available? Besides, Clint is pleased with the idea of using my eggs, because then his store of stem cells will be derived from my eggs, not those of a stranger. We can afford this procedure, and we can afford an attorney!"

INTRODUCTION

Although we have had the technological capability for quite a while to freeze sperm, the technological ability to freeze eggs is a more recent phenomenon. This ability can obviously aid in fertility treatments, as in IVF procedures. But Clint and Marian's situation raises an additional issue: that of being able to use one's own cells and tissues to help regenerate one's aging body parts, including the development of new cells, tissues, and organs to replace damaged ones. It is even possible in the future that we will be able to develop individualized therapy with regard to each person's unique DNA and genetic make-up. These regenerative technologies are the hope and possibility of stem-cell therapies, using both embryonic stem cells as well as adult stem cells. There have been many well-known celebrities, such as Michael J. Fox with Parkinson's disease, and the recently deceased Christopher Reeve with a spinal cord injury, who championed the use of stem-cell therapies. Ethical and religious concerns arise, however, particularly with embryonic stem cells, which have the capacity to develop into virtually any tissue or organ, but which often (but not always) require the destruction of embryos. But even regenerative adult stem-cell therapies raise questions about human lifespan and the possibility of increasing longevity.

We know that scientific advances and technological developments have theological implications, and understanding the science first is very important.[1] It is difficult to separate out the issues of stem-cell research and cloning because similar technologies are used for both. However, we separate out the two issues because although they do raise some overlapping concerns, each technology also raises unique issues. In this chapter we address stem-cell research, and in the next chapter we consider therapeutic and reproductive cloning.

There are three basic issues in the stem-cell debate: should such research be permitted, should it be funded by the government, and does it matter how the embryos are obtained?[2] All of these are important questions, and this chapter focuses particularly on the first and third by exploring some of the more controversial concerns. First, we provide a scientific discussion of embryogenesis, dealing with how embryos are formed, and then describe the issues this raises about when human life begins. Second, we provide the scientific distinctions between embryonic and adult stem-cell research, with a focus on embryos, especially those procedures that harm embryos and which the U.S. government does not fund. Since much of the research on embryonic stem cells does require the destruction of embryos, we address the more common arguments used for and against this research. Third, we describe the science of embryonic stem-cell research that does not harm embryos. We raise some minor ethical concerns even with this technology, which are not as serious as with those technologies that require the destruction of embryos. Fourth, we discuss the scientific future possibilities of extending the human lifespan, and subsequently raise the theological issues related to the Christian understanding of death and salvation, as well as some of the social and emotional concerns that would arise from people living longer, and possibly even forever. Finally, we offer some concluding comments.

EMBRYOGENESIS AND THE ORIGIN OF LIFE

Cells are the fundamental and continuous components of living organisms. Each new generation of cells begins when a sperm cell and an egg cell from the previous generation unite to form a zygote. The subsequent process of cell division along with cell specialization to form tissues in the embryo is called embryogenesis and, through countless cell divisions, ultimately leads to a multicelled organism. The cells in a multicelled organism behave as if they were a "society of cells." The socialization of children may serve to illustrate the significance of stem cells in development and maintenance of an organism. Stem cells are like children: as they mature and assume particular roles in a family (cells form tissues and organs), they acquire characteristics that enable them to assume specific roles in society (organ systems enable the multicelled organism to function as a whole). Taking the analogy one step further, the child's nature (genetic factors in the cell) interacts with the nurturing the child receives within its environment (genes are turned on and off as they interact with factors both internal and external to the cell). Cells, like members of a society, take on particular roles for collective actions—cells are to an organism what people are to a society. It is generally agreed that the development of an organism from fertilization to adulthood, and even to an eventual natural death, involves gene expression influenced by the environment. Genes are turned on and off to orchestrate specialization of stem cells to form the 200-plus mature cells that make up the human body. Genes can, to a limited extent, be turned on and off in wound healing. For example, if one breaks their spine, the bones and muscles will heal, but the cells that make up the spinal cord tissue are unable to orchestrate the entire process—this is where stem-cell technologies may help someday. Modern-day embryologists owe a great deal to the work of

Nobel Prize winner Hans Spemann. His study of amphibian embryos set the stage for our understanding of embryonic induction, a process where specific parts of an embryo influence the development of tissues and organs. Efforts to identify material causes for this process were and continue today as important research for both scientists and ethicists. Spemann, in his classic 1938 text entitled *Embryonic Development and Induction,* offered the possibility that the potency of parts within the embryo is not simply a result of chemical reactions, but may be a vital process of what he called a "psychical nature." Spemann believed that we should not miss this opportunity to seek answers to better understand this possibility.[3] Today, our understanding of both embryogenesis and genetics has enhanced our understanding of embryonic cell potency and how this type of cell potency could be used to develop cell-based regenerative therapies. These therapies could someday extend life by continuous replacement of tissues and organs without danger of rejection. To understand these life-extending therapies, we must first understand how embryos are formed.

Embryogenesis begins with fertilization (normally in the fallopian tubes) and ends when the fetal stage begins at about 8 to 9 weeks into pregnancy (within the uterus). The term embryo is frequently used to describe the three stages that precede the fetal stage: the zygote, blastocyst, and embryonic stages. It is important to distinguish these stages for the purpose of understanding ethical issues associated with the present and future implications of stem-cell technologies. The zygote stage of embryogenesis begins when a sperm (containing 23 paternal chromosomes) is absorbed by the egg (containing 23 maternal chromosomes) to form the zygote (a cell with 46 chromosomes). The zygote then divides to form two cells, then four cells, and then eight identical cells—it takes about 3 to 4 days to reach the eight-celled stage. These cells are called blastomere cells, and they are described as totipotent because they have the potential to form any type of cell in the adult body, including those cells necessary to form the placenta. The blastomere cells continue dividing and, in about 4 to 6 days, they form what is typically called a blastocyst. This is a ball-shaped structure with an outer layer of cells called the trophectoderm surrounding a fluid-filled space that contains an inner mass of about 30 cells pushed to one side. This inner mass of cells continues to divide as the blastocyst travels from the fallopian tubes into the uterus, and if conditions are normal, the trophectoderm of the blastocyst facilitates connection of the blastocyst to the uterine wall—a process called implantation. The blastocyst is commonly referred to as an embryo, or a preimplantation embryo, even though up until implantation, it is by definition a blastocyst. The trophectoderm of the blastocyst is essential for implantation. Trophectoderm cells give rise to chorionic cells that will, on implantation, release human chorionic gonadotropin hormone (tested for in pregnancy kits) to maintain the blood-enriched uterine lining into which fingerlike projections (chorionic villi) will extend to integrate with cells of the uterine wall and ultimately form the placenta. Cells within the inner cell mass of the blastocyst are called pluripotent because they have the potential to become any of the three germ layers—the endoderm, mesoderm, and ectoderm. These layers form on implantation. Pluripotent cells are more limited than totipotent cells because they cannot form the trophectoderm necessary for implantation.

Successful implantation of the blastocyst marks the beginning of the embryonic stage at about 12 to 14 days. The three germ layers can then form with maternal support. The inner cell mass (about 40 to 100 cells) folds in on itself to form the three germ layers in a process called gastrulation. Each germ layer will ultimately give rise to an array of different tissues and organs in specific regions of the body. For example, cells of the endoderm (inner layer) give rise to the inner tissues forming the respiratory and digestive systems. The mesoderm (middle layer) will give rise to tissues forming the skeletal, muscle, and circulatory systems. The ectoderm (outer layer) will give rise to the epidermis, nervous system, and lens of the eye. Cells in each of these germ layers are called multipotent because they are limited to forming tissues within their respective germ layers. Thus, multipotent cells arise through gastrulation on implantation. Ted Peters highlights two ethically important events on implantation of the embryo. First, this embryo is now an individual. It is no longer possible for the embryo to split to produce identical twins. Second, the umbilical connection now allows for the mother's hormones to influence development of a human by acting on genes within the embryo—no hormones, no development. Thus, according to Peters, a blastocyst must be capable of implantation if it is to develop into a human.[4]

But not everyone agrees with Peters! Although scientifically distinctions are made among the zygote, blastocyst, and embryo, in theological and ethical discussions about the beginning of life, the term embryo is generally used to refer to the different stages before the fetal stage. A distinction is sometimes made, however, about the age or development of the embryo, in particular whether or not it has implanted into the uterus. In any case, any technology involving embryos must of necessity ask the question: when does human life begin? The answers do vary, even among theologians and ethicists. All would agree that once human life exists, some kind of protection must exist for this life. Whether one believes the embryo at any stage is a bunch of cells, a potential human life, or an actual human life affects one's view about what kinds of measures should exist to protect it. We now explore some of the possible positions with regard to when human life begins.

In the Christian tradition, conception has usually been considered the point of demarcation for when human life begins. Most people who maintain this realize that conception is not really a moment as much as it is a process, but that since there is no precise moment, we should presume that life begins once the sperm and egg unite. It is at this time that the unique genetic individual exists. It is then a human life, and to destroy it is to destroy a human being, which is a serious moral wrong or sin.[5] This position is held by the Catholic Church and many conservative Christian traditions, including evangelicals. For these groups, then, using embryos at any stage is wrong, and they tend to be unilaterally opposed to embryonic stem-cell research (some individuals may make an exception if the embryos are truly left over and are going to be destroyed anyway). Embryos also represent a vulnerable population, like the seriously disabled who, because they cannot advocate for themselves, should be afforded special protection.

In the past approximately 30 years, however, some theologians have challenged the fertilization-conception line as being definitive. They argue that implantation, which occurs at 12 to 14 days, is the key point at which life begins. The main argument for this position is that before implantation, it is still possible for one embryo to split into two (thus creating identical twins) or for two embryos to unite to become one. Since we do not know even how many individuals exist, we cannot talk about life existing until the embryos implant. Another argument, which is made by Ronald M. Green, is that two thirds to three quarters of all fertilized eggs do not implant in the womb, with many sloughed off which we do not even know about: "In view of this high rate of embryo loss, do we truly want to bestow much moral significance on an entity with which nature is so wasteful? What would be the costs of doing so?"[6] Finally, some consider it important that it is at the point of implantation that the mother's hormones begin to influence fetal development by acting on its genes.[7] Because embryonic stem-cell research occurs well before implantation, these individuals would maintain that using embryos at this stage of development is not problematic.

Though conception and implantation are the most commonly argued for boundary lines, other have argued for life beginning at other points of embryonic and fetal development, such as the presence of brain activity, development of major organs, quickening (when the mother can feel movement within), viability (the point at which the fetus, if born, could exist outside of the mother's womb; with current technology this is usually at about 6 months), and even birth. With regard to the view that biological development is a process rather than a point, one theologian notes: "The trouble is that our biological development is subtle and gradual, but morality wants clarity and sometimes forces it where biology does not permit us to find it."[8] Obviously it is this very clarity that is lacking, at least for many! The issue of when life begins is necessarily related to the moral status of the embryo, which is a bit more complicated than trying to decide when life begins, because even if we could all agree on the beginning point of life, it would not solve some subsequent ethical concerns. In the next section, we describe stem-cell research overall, embryonic stem-cell research that undoubtedly results in the destruction of embryos, and the major arguments on both sides regarding use and ultimate destruction of embryos in research.

STEM-CELL RESEARCH

This section deals with an overview of stem-cell research and embryonic stem-cell research that harms embryos—a later section considers alternatives to harming the embryo. Stem cells differ from ordinary mature body cells in that they have the ability to self-renew; that is, they can divide over and over again without becoming specialized. Stem cells are essential to the development and maintenance of multicelled organisms. In these organisms, every cell nucleus contains all of the genetic information necessary to develop the entire organism, but not all of that information is readily available all of the time. Genetic information can be turned on (DNA is made available for expression) and off (DNA is made unavailable for expression) by various mechanisms in the cell nucleus, and in some cases the genes are turned off indefinitely as cells specialize to form tissues and organs during the process of embryogenesis.

Stem cells are generally described according to two categories: embryonic stem cells and adult stem cells. Both stem-cell types are considered to be sources for cell-based regenerative therapies, but they have significant differences in their natural potential to form a range of cell and tissue types. Embryonic stem (ES) cells are categorized according to the extent of their potency—totipotent or pluripotent. Totipotent ES cells have the potential to become any cell type, including tissue that becomes part of the placenta. Totipotent ES cells include the zygote and its descendent cells (up to and including the eight-cell stage, at about 3 to 4 days). As noted earlier, only these totipotent cells are capable of forming the outer trophectoderm layer that is necessary for implantation to form an embryo. Thus, under natural conditions, only totipotent cells can become an embryo and ultimately implant to form a human. The second category of ES cells is described as pluripotent (about 4 days to implantation and gastrulation). As noted earlier, pluripotent ES cells form the inner cell mass before gastrulation, but not the trophectoderm cells that are necessary to form a viable embryo on implantation. Thus, pluripotent cells cannot naturally develop into an embryo for further development into a fetus. Pluripotent cells can, however, continue to divide in laboratory cell cultures without deterioration, and they have the potential to become any cell in the human body. Pluripotent stem-cell lines that can be maintained in cell cultures where they divide indefinitely are said to be "characterized." Scientists no longer use the word "immortal" to describe these cell lines—this is a theological concept.[9] ES cells are of tremendous therapeutic potential, because when these cells are transplanted into an organism, they have the potential to regenerate all types of complex tissues with a wide range of cell types, and may even someday be used to regenerate entire organs.

Embryonic stem cells are capable of building the entire organism, but adult stem (AS) cells can only maintain that organism. The stem cells discovered in adults do not, under natural conditions, have the potential to develop into any type of tissue. These AS cells are multipotent. To understand why, recall that after implantation the process of gastrulation establishes three embryonic germ layers (the endoderm, mesoderm, and ectoderm). AS cells contained within each germ layer are multipotent because they have specialized to the point of being able to produce cells and tissues within their own germ layer and no other. For example, multipotent blood stem cells (or hematopoietic stem cells) within the mesoderm can become all types of blood cells (platelets, red, and white blood cells), but they cannot form a neuron cell because a neuron cell is derived from the ectodermal germ layer. As cells in each of the three germ layers become more and more specialized, they eventually are dedicated to one type of cell, and under normal conditions, they are no longer able to become another type of cell; they are now called unipotent. These unipotent AS cells are able to renew themselves, but cannot produce another type of cell. Fibroblast skin cells are one example; they continue to renew and maintain connective tissues and they play an important role in maintenance and wound healing. Genes orchestrate ES and AS cell activity in development and maintenance, respectively. ES cells are like musicians in an orchestra. Musicians and their instruments are directed to express their sounds according to a sequence of notes on a staff (compositions). ES cells and their biochemical systems are directed to express their form and function according to chemical base sequences on strands of DNA (genes). Coordination is essential in both cases. When the symphony is finished, the musician closes her music and carries it home to practice and maintain her skills. When development of an organism is finished, the genes expressed to direct ES cells in the orchestration of development are deactivated as they become AS cells and mature cells. The persistence of AS cells in mature tissues may be nature's way of providing back-up cells for body maintenance and repair. For example, many adult tissues, such as bone marrow, intestines, skin, and the cornea, contain AS cells that live alongside specialized cells in the tissue. These AS cells (also called progenitor cells) are capable of renewal and future specialization for the regeneration of tissues damaged by wear and tear.[10]

Embryonic stem-cell research is much more controversial than adult stem-cell research. Many of the political and ethical concerns surrounding stem-cell research are the result of how ES cells are obtained for use in research to develop cell-based regenerative therapies. ES cells are obtained from donated IVF blastocysts (pre-implantation embryos), and they have great value in medical research because they can become any type of cell in the adult body (they are pluripotent). Human embryonic stem (hES) cell lines can be derived from preimplantation embryos produced by IVF procedures. First, the trophectoderm is removed so that blastomere cells can be separated for placement on a laboratory feeder tray. If conditions are right, a colony of pluripotent stem cells will then develop and continue to renew through cell divisions without deterioration (become characterized). The typical feeder tray contains a layer of embryonic skin cells derived from a pregnant mouse to serve as a surface on which the hES cells are nurtured, including regular separation to avoid overcrowding. Recently, new techniques that do not rely on mouse feeder cells have been developed to avoid the risk of viruses and other possible contaminants in mouse cells.[11] With proper care, these cells continue to divide and serve as a source of pluripotent cells for medical research and potential therapies. It is estimated that there are at least 155 hES cell lines worldwide, including the first hES cell lines developed in 1998 by James Thomson at the University of Wisconsin in Madison.[12] Federal funding for hES cell research changed significantly on August 9, 2001, when President Bush announced that only hES cell lines meeting three criteria would receive federal funding. First, all hES cell lines to be used in research must have been derived prior to 9 P.M. EDT, August 9, 2001; second, the embryo must have been created for reproductive purposes and no longer needed (rather than created especially for this research); and third, informed consent must have been obtained, without financial inducements, from the parents of the embryos. The NIH implemented these criteria, and it reported that investigators from 14 laboratories have derived cells from 71 individual, genetically diverse blastocysts that meet the president's criteria.[13] It remains to be seen how long these approved hES cell lines will continue without deterioration. A significant shift in policy occurred on March 9, 2009, when President Barack Obama overturned President Bush's August 9, 2001, announcement. New stemcell lines can now be derived with parental consent using embryos created by IVF procedures for reproductive purposes only and that are no longer needed. The NIH is currently drafting new guidelines to fund human stem cell research.[14] It is not known whether or not private and public funding will be adequate to sustain ES cell research in the United States as other countries push to take a lead in these biotechnologies; Singapore's Biopolis is an example.[15]

Research using embryos is likely to continue, but scientists are seeking ways to develop cell lines without the destruction of embryos.

It is obvious that these early totipotent cells have the most versatility in terms of developing into other body tissues and organs, such as for repair. Thus, embryonic stem cells in this regard are more desirable than adult stem cells. For those who maintain that life begins at conception, though, any use of embryos that results in their destruction is problematic. Several frameworks surround the ethical discussion of the use of embryonic stem-cells: the embryo protection debate, the nature protection debate, the medical benefits debate, and the research standards debate.[16] Most opponents of embryonic stem-cell research argue from the basis of protection of the embryo, while advocates tend to be on the medical benefits side; but in either case, an ethical discussion of the moral status of the embryo is necessary. Obviously, part of this discussion is what has already been addressed—when life begins. The issue of moral status relates to when life deserves protection. Even if one agrees that life exists at the moment of conception, there is the still the additional issue of how this life should fare with regard to other kinds of life in terms of protection and status. To say that all life has dignity and sanctity does not answer the further question of whose life has higher value with regard to a possible conflict of protection, if such a conflictual situation exists. We consider the following issues with regard to arguments about embryonic stem-cell research: deontological versus consequentialist arguments; the question of personhood; the relationship to the abortion issue; the slippery slope argument; and the issue of how embryos are obtained.

Deontological arguments generally assume that certain actions are always wrong, that we can know at least what some of these actions are, and that it is morally wrong to violate these normative principles by engaging in wrong actions. For many conservative Christians, for example, abortion is such an issue, and obviously the moral status of the embryo is at the heart of the abortion debate. God is the creator of life; life is sacred and has dignity; and because this sacred life has been given to us by God, we cannot take this life away (such as in destroying embryos)—regardless of the consequences. Deontological theorists do not make appeals to good consequences, even if they exist, to make their argument; it is the intrinsic nature of the action that is important. If an action is intrinsically wrong (that is, wrong by its very nature), then it is wrong regardless of what benefits it might yield. Thus, destroying embryos is wrong because it destroys life, even if this destruction might help many others. The end cannot justify the means! Consequentialist arguments, on the other hand, argue on the basis of the expected good results. These arguments are based on the idea that, at times, the end can certainly justify the means! In this case, it means that one can still believe in God and even acknowledge that embryos are life, but maintain that these lives can be sacrificed for the greater good, such as might be the case with regenerative therapies resulting from the destruction of embryos. Consequentialists could also argue with regard to embryonic stem-cell research, as well as with adult stem-cell research, that we have obligations to engage in this research for the benefit of future generations.

A common distinction used in medical ethics is the distinction between a human being and a person. In everyday language, we often use these terms interchangeably, but philosophers and medical ethicists use a more precise application to distinguish between those humans who have normal cognitive ability and those who do not. By definition, a human being is one who has human DNA and has been born to human parents. A person, on the other hand, is one who has the cognitive capacity to be autonomous and thus is able to make decisions for oneself. Some criteria for personhood include the ability to reason, to set goals, to have relationships, and to have a sense of self. If, for the sake of argument, one were to allow for this distinction, then it would be possible for some individuals to be humans and persons (all cognitively normal adult humans), for some to be humans but not persons (all humans with cognitive deficiencies, including those with severe dementia and certainly embryos), and for some to be persons but not humans (some higher animals). The attribution of personhood is important because along with it comes rights and the subsequent protection of those rights. It should be pointed out, though, that this distinction between humans and persons is morally repugnant to some Christians who do not want to distinguish among different levels of human beings, especially when it comes to the severely mentally challenged and to embryos. But this distinction can be important in health care when we have to make tough decisions about how resources should be allocated and who should have priority when there are conflicts. Thus, for those who consider embryos to be human, or potentially human, it is still possible for them to maintain that embryos can be destroyed to help other humans who are persons.

The use of embryos in stem-cell research is also directly related to the abortion debate. There is probably a direct correlation between those who are opposed to abortion in virtually all cases (typically the "prolife" view) and those who are opposed to embryonic stem-cell research that destroys embryos. This is because the moral status of the embryo is considered the same—it is just as wrong to abort them as to destroy them in research because in both cases it is the destruction of life. It is important to point out as well, that for those who believe that life begins at conception, they often believe that this is the point when the individual receives a "soul." However, those who argue that abortion can be justified in a number of cases usually maintain this because they think that the mother's rights should ordinarily trump the embryo's rights, even if they think that the embryo is human life or potential life (typically the "prochoice" view). Since abortion is already legal and we are destroying embryos in this arena, they do not think it is necessarily problematic to destroy embryos in research, especially if there is hope that it can help many with regenerative therapies. In addition, prochoice individuals might maintain that we should use discarded embryos from abortions for research so that they will not have been destroyed for nothing.

One of the concerns with regard to research on embryos is the slippery slope argument. This type of argumentation is concerned about the next possible step in a particular course of action. Thus, right now researchers primarily are using preimplantation embryos in stem-cell research. But what if it turns out that embryos that are more developed can be useful—should we be able to use them? What if this would result in the destruction of embryos that would be past the stage of implantation? This could raise concerns even for those who believe that implantation is the morally significant line of demarcation. Those who oppose this slippery slope argument could argue against it in two ways: by pointing out that since we are only dealing with preimplantation embryos right now, we do not need to consider this subsequent possibility, or that if it does become a possibility, it may not be a line that we wish to cross.

How embryos are obtained is important in the embryonic stem-cell debate. Embryos can be obtained in several ways: fresh embryos created from egg and sperm in the laboratory (the kind of procedures now done in IVF techniques); cloned embryos (which raise concerns about the moral desirability of cloning); and excess frozen embryos (for example, embryos already created from previous IVF procedures but not expected to be implanted).[17] Each has its own problems. The most troublesome would be the deliberate creation of embryos simply to be destroyed, and groups such as the Catholic Church would be in opposition because of the IVF technique itself that is used. Cloning embryos raises some of the same kinds of issues. Using excess embryos might be the least controversial, because they are already created and cannot continue to live in perpetuity (or if they can, there is no point to it). In fact, existing stem-cell lines have been created from excess embryos in the past, but this does not necessarily grant permission for using frozen embryos in the future. Obviously, the use of embryos that results in their destruction is morally problematic and an issue on which we may not achieve consensus. If there was a way, however, to employ the versatility of embryonic stem cells without destroying embryos, it would certainly be a resolution that both sides could embrace, or at least live with. We now explore scientific possibilities in this regard, as well as some ethical issues that could arise even with this kind of research.

RESEARCH EFFORTS TO AVOID HARMING EMBRYOS

Several research techniques to develop stem-cell lines can be distinguished by the extent to which the embryo is directly involved in the process. The assumption is that the less an embryo is involved these techniques, the less chance of damage (or harm) to the embryo. Some techniques avoid using embryos altogether; they include AS cell research, and efforts to reprogram adult cells to function like embryonic cells. Some techniques continue to use embryos while making an effort to avoid damaging (or harming) the embryo; they include using embryos that are no longer viable and efforts to obtain cells from embryos without damaging the long-term viability of the embryo. It is likely, however, that embryos have been destroyed in the development of these techniques.

The most obvious way of avoiding any harm to an embryo is to avoid using it altogether in the research and development of cell-based therapies. Two approaches include using AS cells and efforts to reactivate dormant embryonic genes in somatic cells. These cells can be obtained from biopsying adult tissues, and therefore avoid using embryos, but because these cells are multi- and unipotent, their range of application is limited. For example, stem-cell transplantation is an approach sometimes used in an attempt to treat multiple myeloma. This type of cancer affects a specific kind of white blood cell, the plasma stem cell, which is part of our immune system and is produced in the bone marrow.[18] The AS cells necessary for this type of cell-based therapy can be obtained without controversy from blood and bone marrow tissue. Research and development of these therapies depend on the availability of specific AS cells (recall that they are multipotent). Efforts are under way to create pluripotent cells from adult cells. Scientists have long recognized that all somatic cells with a nucleus contain the necessary genetic information to become any cell type, or even to develop into a human clone. But it was generally thought that this genetic information was permanently deactivated as cells became dedicated to their mature form and function.

We now know that it is also possible to reactivate dormant embryonic genes in adult somatic cells. Ian Wilmut and his coworkers demonstrated in 1997 that it is possible to reprogram mature cell nuclei by cloning a sheep named Dolly using a technique called somatic cell nuclear transfer (SCNT). Recall the introductory case in which Marian's plan was to have a nucleus taken from one of her mature body cells (a somatic cell nucleus) and that nucleus was to then be transferred to one of her eggs with the nucleus removed (this is SCNT). Factors remaining in the cytoplasm of her egg would then reactivate deactivated genetic information contained in her somatic cell nucleus. The reprogrammed cell could then divide and provide a source of pluripotent stem cells. An added benefit for Marian is that these cells would be her genetic match and could therefore provide her with a source of compatible replacement tissues.

There are two techniques to reprogram adult cells to function as pluripotent stem (PS) cells: developing induced pluripotent stem (iPS) cells, and developing parthenogenic pluripotent stem (pPS) cells. Neither of these techniques requires using an embryo, but ethical concerns remain because both iPS and pPS cells, in addition to becoming pluripotent, may also become totipotent (capable of forming a human clone if implanted into a uterus). How are these iPS cells created? The iPS approach builds on the well-developed SCNT method. Recall that this technique includes the transfer of a nucleus from a somatic cell to an oocyte (egg) with its nucleus removed for reprogramming to form a PS cell.[19] Whether or not these are embryonic cells, or adult cells that just behave like embryonic cells, is not a relevant scientific question, but may be a question for ethicists. What about concerns that these PS cells are totipotent and thus have the potential to form a human if implanted in a uterus? Is there a scientific solution to this ethical dilemma? Richard Hurlbut, a member of the U.S. President's Council on Bioethics, asserts that the somatic cells used for SCNT could be genetically modified to preclude formation of the trophectoderm, and thus pluripotent stem-cell lines developed in this way would not be capable of becoming human embryos.[20] Scientific methods to preclude trophectoderm formation are not yet developed.

The pPS cell technique employs a process in nature, called parthenogenesis, that occurs when an embryo develops directly from an oocyte without being fertilized. Parthenogenesis is a reproductive process that has been observed to occur naturally in some plants and animals, but not in humans as far as we know. Efforts to induce parthenogenesis in mouse oocytes for the production of pPS cell-lines have been successful. A chemical that inhibits cell division can be used to prevent expulsion of a polar body during oocyte formation.[21] The resulting oocyte is called a pseudozygote, and it can develop into a blastocyst, which can then be used to develop into a pPS cell line. These cells are pluripotent like ES cells, but because they are of maternal origin only, they lack the paternal genes that are essential for normal development (this is an example of male imprinting).[22] This observation may set aside some concerns that pPS stem cells are embryos with the potential to become human, but it remains to be seen if this characteristic will be an impediment to the therapeutic potential of these stem-cell lines. The impetus for developing both pPS and iPS cell techniques is their tremendous therapeutic potential for production of tissues and organs that are a genetic match to the somatic cell donor.

Of the techniques that use embryos but make an effort to avoid damaging (or harming) them, the first and most obvious technique is to use embryos that are no longer viable—so-called "dead" embryos. This approach assumes that no harm can be done to the embryo because it is dead in its capacity to form an organism. This approach has been suggested by the President's Council on Bioethics, and it is generally considered to be morally acceptable because it is similar to organ donation by a person who is judged to be "brain dead." The outcomes of this research are uncertain. Can we determine when a blastocyst is no longer viable? How useful are these so-called dead embryos and their cells?[23]

A second method is to remove one or more cells from an embryo without affecting the viability of the embryo; it is called single blastomere biopsy (SBB). The procedure begins with the removal of a single blastomere cell from a blastocyst; this procedure is already used to detect genetic defects in the PGD of embryos. SBB does not require the destruction of embryos. It has been used in mice to demonstrate that it is possible to develop ES cell lines without harm to the parent embryo and its ability to develop.[24] Advanced Cell Technology (ACT)[25] is a U.S. biotechnology company and a leader in the efforts to develop cell-based therapies that avoid destruction of human embryos. The work of ACT indicates that a single biopsied blastomere cell, removed for genetic testing during a clinical PGD procedure, can be grown overnight to yield multiple cells for both genetic testing and the derivation of pluripotent human ES cell lines.[26] Thus, it may be possible to prepare pluripotent human ES cell lines without harm to donor embryos. It remains to be demonstrated with certainty that hES cell lines derived using these methods will not form a trophectoderm layer—and thus a human embryo.

Obviously, the moral concerns raised with regard to embryonic stem-cell research that does not harm embryos are not as serious as those related to the questions of the life and subsequent moral status of embryos. In addition, not a lot of attention has been paid to them, ethically speaking. But a few minor concerns can be raised if we arrive at a point when we can create embryonic stem cells without harming embryos. Permissible research would need to include several factors. Consent must be obtained from those responsible for the creation of the embryo that might be used in research, even if it is not harmed. Disproportionate risks for the embryo and the mother must be avoided in any technology developed. A legitimate hope of benefit must exist for others.[27] We can also raise some hypothetical questions. If any embryos were destroyed to create these genetic techniques, does this override the use of embryos in research at all? What about the possibility that a blastocyst (a so-called dead embryo) is viable? In SBB, is there immediate damage or long-term negative consequences? What is the relationship of the use of embryos and IVF procedures in general? Until we can adequately answer some of these questions and concerns, it is likely that embryonic stem-cell research will be problematic, but certainly much less so if we can be sure that embryos are not being harmed.

The use of adult stem cells, however, is the least controversial of all, and their possibilities for regenerative medicine are significant. We have already described the regenerative techniques for which stem-cell research offers hope, so we turn our attention briefly to the more fanciful notion of living forever.

LIVING FOREVER: FACT OR FANCY?

The science has a way to go, but someday it is very likely, because of stem-cell research today, that we will be able to extend our lives by reprogramming genetic material within our cell nuclei to develop a constant supply (in vitro) of personal replacement cells, tissues, and even organs as needed. We may someday be able to turn genes on and off within specific cells in our bodies (in vivo) to maintain and regenerate cells, tissues, and organs as needed for a very long life. Ronald M. Green reminds us that none of these stem-cell technologies is without scientific, ethical, and political challenges. SBB and pES techniques are probably closest to clinical application, but a breakthrough in somatic cell dedifferentiation research would significantly influence the discussion.[28] Technologies to produce iPS cells from somatic cells may be the kind of breakthrough technology that Green is talking about. A team of researchers at Columbia and Harvard Universities successfully reprogrammed human skin fibroblast cells (cells common in connective tissue) from an 82-year-old woman with a familial form of amyotrophic lateral sclerosis (ALS), commonly called Lou Gehrig's disease. The team used virus vectors to introduce four genes that then induced the skin fibroblast cells to dedifferentiate in tissue culture. These reprogrammed cells are pluripotent because they are capable of forming tissues in any one of the three germ layers, but in this case, the researchers cultured terminally differentiated nervous tissue. This work is of great significance because it demonstrates that human iPS cell lines can be derived using somatic cells from older adults and from diseased tissues. These cell lines can serve as models when studying a disease at the cellular level, and they could ultimately produce genetically matched tissues for regenerative therapies. It is unlikely that this approach will move to the clinical stage until the genes with a potential to cause cancer and retrovirus vectors used in this technique are controlled or replaced.[29]

Christianity already believes in the possibility of immortality—that ideally we will live with God forever in the next life. But the thought that we could live forever in this life challenges some traditional theological concepts. Using stem cells to repair organs and tissues if it does not harm other life forms is not problematic, and it is likely that this kind of therapy will be more common in the future. But what if we are eventually able to significantly extend our lifespan greatly beyond what it is today, and what if we were able to get to a point (certainly now only within the province of science fiction) to live forever? What would that mean for us as a species: socially, medically, psychologically, cognitively, and theologically? Obviously, living forever challenges the Christian doctrines of sin and the Fall, death and salvation. Christianity has traditionally maintained that although death is contrary to the original and perhaps ultimate purpose of God for humans, it is a result of our sinful condition. Christianity also maintains that the death of Jesus is redemptive and that there is an afterlife, which living forever might challenge. Paul Ramsey, a well-known Christian ethicist, said many years ago when addressing issues in genetics: "Religious people have never denied, indeed they affirm, that God means to kill us all in the end, and in the end He is going to succeed."[30] Whether we think that Ramsey has overstated the case or not, living forever in our physical bodies on the planet Earth is certainly an idea that is foreign to Christianity!

However, the search for immorality has been around for a long time. Within biomedicine, we can identify four strategies that biomedicine has used to handle the knowledge of aging and death. The first strategy is to normalize aging. This is demonstrated by the fact that we have attempted to help individuals avoid premature death, especially through the significant increase in the human lifespan over the past 150 years. The second strategy, and where medicine currently is, is to optimize aging, so that it is no longer viewed as an inevitable and natural process. The effort now is to mitigate age-related diseases, and to argue that the current lifespan of 75–80 years should be viewed as an arbitrary boundary that need not be inelastic. The third strategy is to retard aging and postpone death; one scientist even has argued for the eventual possibility of and even desirability of a 1,000-year lifespan! The fourth strategy is to eliminate or overcome aging and death; it is here where terms such as "transhuman" and "posthuman" are used.[31] These terms raise the idea of a state or condition where we move beyond being "human" and are no longer bound by our conceptions of what this means.

There are other concerns with extended life as well. We will not know in advance the side effects of genetic changes that may be used to extend life. Thus, we probably should proceed cautiously where risks are high. There would be considerable demographic problems if people lived hundreds of years. Living forever tends to focus on the individual rather than on relational ideas of what it means to be human.[32] Christian perfection has tended to focus on moral perfection rather than on the physical perfection of a body that never or only very slowly ages. If a moment of death is no longer tenable, then what happens to the idea of a soul and the afterlife?[33] As we can see, the concept of increased aging as well as living forever raises some problematic issues, but they are certainly ones that are not likely to be a problem in the very near future.

CONCLUSION

The issue of stem-cell research, or therapeutic cloning, is a very complicated one. Oftentimes people ask others: Do you believe in or support stem-cell research? But as we have seen, the possible answers must be much more nuanced than simply "yes" or "no." Many individuals do not even have the basic knowledge about how embryos are formed, let alone stem-cell lines. With regard to this research, we must ask: Are we talking about adult stem cells or embryonic stem cells? Are we talking about techniques involving harm to embryos, or would those techniques that do not harm embryos belong in a different category? It is likely that these ethical concerns regarding stem-cell research would largely go away if we are eventually able to either create embryonic-like versatility in adult stem cells or if we could perfect techniques that would never harm embryos. But as long as we must continue to think about embryos, we cannot escape questions such as: When does life begin? What kind of protection should this life have? When in opposition, whose rights should trump whose, the embryos or those who would benefit from regenerative possibilities? Does it matter from whence the embryos are derived? Obviously, using only adult stem cells will mitigate some of these concerns. But even if we were all to agree on when life begins and on the moral status of the embryo, it still does not answer the subsequent questions of how far we should go in extending life, whether immortality is a life worth choosing, and how to best mitigate the results likely to come. In addition to the controversial issue of therapeutic cloning, we now move on to perhaps the even more controversial notion of reproductive cloning.

NOTES

1. For an excellent resource on different aspects of genetic engineering, including stem-cell research, see Barbara Wexler, *Genetics and Genetic Engineering* (Detroit and New York: Thomson/Gale, 2008).

2. Michael J. Sandel, *The Case against Perfection: Ethics in the Age of Genetic Engineering* (Cambridge, MA, and London: Belknap Press of Harvard University Press, 2007), 102–4. This is a concise yet excellent book on concerns with efforts for genetic perfection.

3. Hans Spemann, *Embryonic Development and Induction* (New Haven, CT: Yale University Press, 1938), 367–72.

4. Ted Peters, *The Stem Cell Debate* (Minneapolis: Fortress Press, 2007), 7–9.

5. A good volume representing the Catholic positions on these issues is John F. Morris, ed., *Medicine, Health Care, and Ethics: Catholic Voices* (Washington, DC: Catholic University Press of America, 2007).

6. Ronald M. Green, *The Human Embryo Research Debates: Bioethics in the Vortex of Controversy* (Oxford and New York: Oxford University Press, 2001), 37.

7. Peters, *The Stem Cell Debate*, 7–9.

8. Ronald Cole-Turner, "Principles and Politics: Beyond the Impasse over the Embryo," in *God and the Embryo: Religious Voices on Stem Cells and Cloning*, edited by Brent Waters and Ronald Cole-Turner (Washington, DC: Georgetown University Press, 2003), 90.

9. Peters, *The Stem Cell Debate*, 1–6.

10. Not all organisms exhibit the same ability to regenerate. For example, a starfish can regenerate an arm, but humans cannot.

11. The following source is very informative: http://stemcells.nih.gov/info/basics/basics3.asp (accessed September 16, 2008).

12. Rick Weiss and Max Aguilera-Hellweg, "The Stem-Cell Divide," *National Geographic* 208 (July 2005): 2–27.

13. http://stemcells.nih.gov/policy (accessed September 17, 2008).

14. http://stemcells.nih.gov/policy/2009draft.htm (accessed April 25, 2009).

15. Visit Biopolis at http://www.one-north.sg/hubs_biopolis.aspx (accessed October 4, 2008).

16. Peters, *The Stem Cell Debate*, 24–28.

17. Rose M. Morgan, *The Genetics Revolution: History, Fears, and Future of a Life-Altering Science* (Westport, CT, and London: Greenwood Press, 2006), 136–37.

18. For further information on this form of cancer, see MayoClinic.com at http://www.mayoclinic.com/health/multiple-myeloma/DS00415 (accessed on October 4, 2008).

19. These cells are chimeras in that they have the nuclear DNA of the somatic cell and the mitochondrial DNA of the oocyte. If an oocyte from another mammal is used, then this is a transgenic cell—that is, it contains genes from two different species. For more details, visit http://www. bioethics.ac.uk/index.php?do=topic&sid=13 (accessed September 20, 2008).

20. James F. Battey, Chair, NIH Stem Cell Task Force, and Director, National Institute on Deafness and Other Communication Disorders, "Senate Testimony: Alternative Methods of Obtaining Embryonic Stem Cells," in *Testimony Before the Subcommittee on Labor, Health and Human Services, Education, and Related Agencies*, Washington, DC, July 12, 2005, http://stemcells.nih.gov/policy/statements/20050712battey.asp (accessed September 17, 2008).

21. Meiotic cell divisions produce four haploid gamete nuclei. Specifically in females, only one of the four nuclei goes on to form an oocyte, and it is much larger than the remaining nuclei because it contains materials necessary for development. The remaining nuclei become a byproduct of meiosis and are called polar bodies. In males, the four cell nuclei are similar in size, and they each go on to form sperm cells.

22. Kim Kitai et al., "Histocompatible Embryonic Stem Cells by Parthenogenesis," *Science* 315, no. 5811 (January 26, 2007): 482–86.

23. Battey, "Senate Testimony: Alternative Methods."

24. Young Chung et al., "Embryonic and Extraembryonic Stem Cell Lines Derived from Single Mouse Blastomeres," *Nature* 439 (January 12, 2006): 216–19.

25. http://www.advancedcell.com (accessed September 20, 2008).

26. Irina Klimanskaya et al., "Human Embryonic Stem Cell Lines Derived from Single Blastomeres," *Nature* 444 (November 23, 2006): 481–85.

27. Thomas A. Shannon, "The Roman Catholic Magisterium and Genetic Research: An Overview and Evaluation," in *Design and Destiny: Jewish and Christian Perspectives on Human Germline Modification*, ed. Ronald Cole-Turner (Cambridge, MA, and London: MIT Press, 2008), 55–56.

28. Ronald M. Green, "Can We Develop Ethically Universal Embryonic Stem-Cell Lines?" *Nature Reviews Genetics* 8 (June 2007): 480–85.

29. John T. Dimos et al., "Induced Pluripotent Stem Cells Generated from Patients with ALS Can Be Differentiated into Motor Neurons," *Science* 321, no. 5893 (August 29, 2008): 1218–21.

30. Paul Ramsey, *Fabricated Man: The Ethics of Genetic Control* (New Haven, CT, and London: Yale University Press, 1970), 27.

31. This section has come from the excellent article on immorality by Ulf Görman, "Never Too Late to Live a Little Longer? The Quest for

Extended Life and Immortality—Some Ethical Considerations," in *Future Perfect? God, Medicine, and Human Identity*, ed. Celia Deane-Drummond and Peter Manley Scott (London and New York: T&T Clark International, 2006), 143–54.

32. Ibid., 150–52.

33. Celia Deane-Drummond, "Future Perfect? God, the Transhuman Future and the Quest for Immortality," in *Future Perfect? God, Medicine, and Human Identity*, ed. Celia Deane-Drummond and Peter Manley Scott (London and New York: T&T Clark International, 2006), 172–74.

6

To Duplicate or Not to Duplicate: The Question of Cloning

This is the third time this week that Jimmy's parents were asked to visit the school counselor. Somehow the students in Jimmy's biology class figured out that his older brother, Allen, was his identical twin, and his classmates were mercilessly ridiculing him. "What is going on here?" the counselor asked, but she was not prepared for the truth. Jimmy's parents, Joe and Kate, told her that many years ago, they participated in a privately funded experimental study at a fertility clinic. As part of their IVF procedure, Joe and Kate agreed to allow their doctors to split several of their embryos before the eight-celled stage—the embryos then developed into normal blastocysts. But when Joe and Kate heard that embryos split naturally in nature to yield identical twins, they requested that no blastocysts with the same genetic make-up be inserted into Kate's uterus; this was because they did not want identical twins at this time. A baby boy named Allen was born, and the extra embryos were stored. Several years later, Allen developed an infection that severely damaged his kidneys. It was not going to be easy to find a matching donor for Allen, and a lifetime of antirejection drugs or kidney dialysis were not the options Joe and Kate wanted for their son. They chose instead to screen their stored blastocysts for a genetic match to Allen. The chances of success were slim, but doctors were able to employ the splitting technique to clone more embryos and Jimmy was born—he was Allen's identical twin but was several years younger. Jimmy eventually became a kidney donor for Allen. The counselor's jaw dropped when she heard this story, and after a long pause, she spoke directly to Joe and Kate. "How do you think it feels to be the back-up clone?" Jimmy interrupted, "I am not a back-up clone! I am Allen's identical twin. And it is my kidney that keeps him alive. I love my brother, but ever since learning about clones in biology class, everyone looks at me as if I am just another Allen." Joe and Kate were at a loss for words—what should we do, they wondered.

INTRODUCTION

A "doppelganger" is a German word for the "ghostly counterpart of a living person."[1] This is the kind of image that is often conjured up when we hear the term "clone"—the idea that there may be another copy of us somewhere out there in the world, perhaps occurring naturally. But of even more concern are those eerie possibilities raised by movies such as *The Boys from Brazil,* in which clone-like creatures of Hitler are artificially created with the potential to wreak havoc on the world.[2] No doubt these kinds of fantasies are fueled by science fiction, and yet we are now in a situation where we can not only clone cells and organs, but also are able to create an entire creature from the cell of another. Of course, this has only successfully been done in animals. The very thought of a human clone seems to violate our sense of uniqueness, and it invokes concerns that the natural order of things will be violated if we clone humans. In terms of "playing God," this really seems to push the boundaries of what limits humans should have.

But why would we even want to create a human clone? It is indeed hard to find good reasons for this, and because of this (as well as the inherent dangers of this technology), many oppose human cloning. However, the example we opened this chapter with indicates one reason why we might want to someday do this: to create an exact genetic match in terms of organs for another person. This naturally occurs in twins, but the question remains as to whether or not we should artificially create clones for this purpose. Did Joe and Kate do the right thing in even creating a "back-up" child for Allen? What effect might this have on Jimmy's identity as an individual— will he ever be able to consider himself a separate individual, knowing that he was ultimately created to support Allen's life? Are there possible dangers to this technology of which we are not yet aware— unintended consequences that may be detrimental to Jimmy?[3] The questions are many and the answers are fuzzy.

In this chapter we examine the process of cloning as well as some of the ethical and legislative concerns raised by it. First, we describe how cloning occurs as a natural process in nature and maintain that from an ethical perspective, this is not especially controversial. We then move onto cloning technologies in the laboratory, which raises the possibility of cloning of humans. Because it is with the latter that we have many concerns as a species, we next look at some of the ethical concerns raised by this possibility, and give a brief over-view of legislation on cloning as it currently exists in the United States. Finally, we offer some concluding comments. Stem-cell research does employ cloning techniques that are collectively referred to as therapeutic cloning, but in this chapter we focus on cloning for reproductive purposes.[4]

CLONING IN NATURE

For many of us, the term cloning has a negative connotation. It is, however, a natural occurrence in nature. The ambiguity surround-ing the term clone can be confusing, but biologists generally con-sider clones to be copies of macromolecules such as DNA, and genetic copies of both cells and organisms. We first discuss identical twins as an example of spontaneous human cloning. We then look at cloning as a form of asexual reproduction in nature where both cells and organisms form clones on a regular basis as an essential process for the cycle of life on earth. We then consider cloning at the cellular level where asexual cell division (mitosis) plays an im-portant role in the development and maintenance of organisms. Finally, we point to the need for genetic diversity as a reason not to clone humans.

The spontaneous formation of identical twins is a clear example of natural cloning in humans. Twins are, however, genetic clones of each other; they are not clones of either parent. This is because sex-ual reproduction forms the zygote, which then splits to form twins or clones. Recall that in humans, the cell division that produces game-tes is meiosis. This cell division process duplicates, or clones, the DNA, and then distributes it so that only one-half of the genetic complement (23 chromosomes in humans) is placed in each sperm or egg. The full genetic complement (46 chromosomes) is then restored upon fertilization when sperm meets egg. The entire process is called sexual reproduction. Thus sexual reproduction mixes cloned paternal and maternal genes (carried in the sperm and egg, respec-tively) to produce a genetically unique zygote with the potential for development into an individual person. Sometimes during female ovulation more than one egg is released. Fertilization of each egg may then result two or more genetically unique zygotes and subsequent multiple births to produce fraternal twins, about 12 pairs per 1,000 natural births—these are not clones. But on rare occasions, about 4 in 1,000 births, the fertilized egg spontaneously splits to form identical twins, and each child may under normal circumstances develop in different amniotic sacs while sharing the same placenta.[5] Identical twins are genetic clones, yet we do not commonly refer to these twins as clones. We appreciate their similarities and acquired differences—even fingerprints differentiate to some extent, possibly through tactile differences in the womb. Identical twins are unique to the extent that the environment can influence the expression of their genes. Thus, human clones would not be doppelgangers—not even close—because identical twins growing up together probably do not view the other as a ghostly image. Meiosis is the cell division process that produces gametes to recombine and produce a genetically unique individual; even identical twins are unique relative to their parents. In contrast, mitosis is the cell division process that produces cloned body (somatic) cells. An important difference is that meiosis in sexual reproduction introduces genetic variability: mitosis in asexual reproduction conserves genetic information. Mitosis is the predominant cell division process in reproduction without sex.

Biologists call reproduction without sex asexual reproduction. This type of reproduction can produce genetically identical cells and organisms with a high degree of reliability; these are clones. This type of reproduction is, as the name implies, best described by what it is not. Asexual reproduction is any form of reproduction that does not include meiosis and immediate subsequent fertilization. One clear example is vegetative propagation in plants, such as the rooting of plant cuttings. The new plant (the rooted cutting) is a result of mitotic cell divisions in cells from the original plant. If this were possible in humans, we could grow an entirely new "you" from one of your body parts; more realistically this could be done by nuclear transfer, which we consider later. Another form of asexual reproduction is parthenogenesis. This natural process may someday be applied in humans to develop therapeutic clones. Parthenogenesis in animals (called apomixes in plants) occurs when an unfertilized egg develops into an adult. Parthenogenesis has been observed to occur naturally in many animals such as insects and reptiles. If parthenogenesis were to occur naturally in human females, and as far as we know it does not, the child (or parthenogen) would be female. Parthenogenesis differs from most forms of asexual reproduction in that offspring are derived from irregular events in egg formation during which a mother's heterozygous genetic information may or may not be shuffled; thus parthenogens are not necessarily exact clones. It is generally agreed that asexual reproduction to produce clones may be an efficient reproductive strategy when the environment is stable or a mate is unavailable. When one is living under these conditions, nature's long-term need to maintain genetic diversity of the species may become a second priority. In sand dollars, for example, the larvae have an ability to split and form clones to rapidly increase in numbers when resources are available, and, as a recent study indicates, they also do this when faced with danger. It is thought that sand dollars can detect the mucus of predatory fish, and then split to become smaller targets.[6] Thus, cloning is an essential process in the cycle of life on earth, but cloning at the cellular level also plays an important role.

Cloning at the cellular level is essential for developing and maintaining the organism. Therapeutic cloning techniques are based on this natural ability of cells to form clones while at the same time preserving their ability to form diverse cell types (we discussed this in the last chapter). Human development viewed at the cellular level illustrates this point. Mitotic cell divisions faithfully duplicated all the genetic information contained in your zygote to ultimately form a multicelled "you." All of your cells with a nucleus are genetically identical, and they have matured to take on specific roles to form the cells, tissues, organs, and organ systems that make up your body. The genes contained in cloned cells (and cloned organisms) are expressed as needed over time and in concert with the environment. To illustrate this point, consider the fact that the blastomere cells of the pre-embryo are clones of the original zygote, and as embryogenesis proceeds, their descendent cells (clones of clones) are influenced by each other and by chemical signals within the tissues. All of the genes are present in each cell, but gene expression varies so much that each cell matures to take on a specific form and function; a mature nerve cell looks and acts differently than a mature liver cell. Thus cloned cells, and their ability to specialize, help ensure that the organism can function consistently in both its external and its internal interactions with the environment. For example, the cells that form your retina respond to light and then deliver reliable signals to cells in your optic nerve; this is an important first step in the integration of both external and internal cellular processes for your survival. Thus an organism is a population of cloned cells that develop and function efficiently in their respective environments. This collective action by cells is essential for survival of the organism. But for a population of organisms, or a species as a whole, successful adaptations require genetic diversity among its members.

Recall that sexual reproduction shuffles the genetic material with each generation to facilitate variability, a source of genetic diversity in a population. Asexual reproduction ensures adequate numbers of cells with reasonably accurate copies of information to build and maintain each organism within the population. Why not skip the shuffle for variability and simply distribute the genetic information to each generation as clones? It is generally agreed that sexual reproduction is the important source of variation within a species. The fact that a diploid body type with sexual reproduction is predominant in so-called higher animals and plants is evidence of this. Variability is produced during gamete formation (meiosis) and through the transmission of gene mutations (spontaneous or environmentally induced) that may be useful, may be detrimental, or may have no effect in the next generation. Variability is further enhanced when one considers all of the potential opportunities for mates to share sperm and eggs (although the process of mate selection is seldom random). In contrast, asexual reproduction limits variability in a species, except for mutation or the occasional exchange of genetic information by various means. This is limited variability because mitotic cell divisions that begin with the zygote and continue through adulthood yield cells with a high degree of genetic fidelity—genetically cloned cells. It is generally agreed among biologists that reproduction without sex came before reproduction with sex. We can imagine the first forms of life as single cells producing clones without a significant mixing of genetic information, only to give way to the powerful forces of natural selection on the varied forms ultimately produced by sexual reproduction. Variability is essential for continued survival of a species as the environment changes. For example, factory soot filled the air and darkened tree trunks during the industrialization of England. A species of predominantly white moths then became easy prey for birds, but the species did not go extinct. That is because a less common and darker version of the same species was no longer easy prey and therefore grew in numbers—the species survived. The importance of variability in a species should not be understated.

Should the need for variability cause us to reconsider the cloning of humans? From the perspective of nature, there may be adequate reason to proceed with caution. The faithful reproduction of cells with a high degree of fidelity is necessary if organisms are to continue generation after generation, but the overall reproductive strategy for a species must also maintain genetic diversity as a source of new forms for adaptation to changing environments. Inbreeding within families and small groups will reduce the viability of offspring. If it is frowned on when cousins have children, it is hard to imagine what our reaction might be to the cloning of humans. Code 46 is a popular science fiction film that portrays a future world where humans lost genetic diversity through an overreliance on IVF and cloning techniques. Code 46 represents a policy that was developed to mitigate the loss of genetic diversity by taking steps to prevent related individuals from having children.[7] To clone or not to clone our species is a question that warrants serious consideration, but in either case we should seek ways to maintain our genetic diversity if we are to adapt to an environment that is more than likely to change over time.

The ethical and theological concerns with cloning in nature are virtually nonexistent. Cloning occurs naturally in plant and animal species through asexual reproduction and in humans through twinning. Three minor theological points can be made about evolution, normal human cloning, and human reproduction. First, the benefit of looking to nature for guidelines on cloning will vary depending on whether one accepts or rejects the theory of evolution. For biologists and for many mainstream Christians, evolution is not a controversial concept; it is simply an explanation that we now accept for how life came into existence in its current form. Of course, for Christians, God needs to be involved in the process somehow, even if God simply set evolution in motion at the beginning, and as long as it retains some concept of the idea of humans being created in the image of God. But for those largely conservative Christians who consider evolution to be contrary to the biblical account of creation, any attempt to draw conclusions about human cloning by using other species as models would be considered wrong-headed. Second, the noncontroversial reality of normal human twins could lead Christians to maintain that the occurrence of identical twins that are truly genetic clones is part of God's plan for humans. It might not be such a leap for some, then, to maintain that other types of cloning might be permissible as well, depending of course, on the reason for their creation. However, an important third point needs to be made about the Christian understanding of sexual reproduction. Christianity has traditionally held that human sexual reproduction must be within the confines of marriage, a sacred covenant similar to that which God has with humans. Sex should occur between a man and a woman, and one important purpose of sexuality is reproduction. There have been challenges to this traditional understanding, including whether sex needs to be confined to marriage, whether sex or marriage needs to be confined to heterosexuals, and whether reproduction should occur only naturally through the sexual act as opposed to with the assistance of reproductive technologies. But certainly even for those who accept some of these technologies, such as IVF, the idea of creating humans that do not require two parents is controversial because it abandons the traditional Christian understanding of marriage and the family. The feminist utopian novel, *Herland*, describes a society of all women, in which sexual reproduction occurs through parthenogenesis.[8] Some men are given a tour of this utopia, and they do not see the obvious appeal! The fact that the novel ends with a welcoming back of men into the community indicates that many people would consider a world without men and without sexual reproduction—where men would be truly obsolete—to be a world not worth living in.

CLONING IN THE LABORATORY

Ian Wilmut shocked the world in 1997 when his research team announced the successful use of SCNT technology to clone a sheep named Dolly.[9] We understand the cloning of cells in nature, where this process is essential for the development and maintenance of organisms. We understand that cloning is sometimes a reproductive strategy for organisms in the natural world. But why would anyone want to clone cells and organisms in the laboratory? Scientists employ nature's ability to form clones because it is believed that cloning techniques are a key to developing therapies with a tremendous potential to improve the human condition. Cloning methods used in biomedical research to create and develop cell-based therapies are frequently called therapeutic cloning, a topic closely related to stem-cell research we discussed in the last chapter. In contrast, cloning methods used in research to clone children (and other organisms) are frequently called reproductive cloning.

Therapeutic Cloning

The President's Council on Bioethics describes therapeutic cloning for biomedical research as the "production of a cloned human embryo, formed for the (proximate) purpose of using it in research or for extracting its stem cells, with the (ultimate) goals of gaining scientific knowledge of normal and abnormal development and of developing cures for human diseases."[10] A primary goal for this research is the development of regenerative therapies with the potential to heal or replace all types of cells, tissues, and even organs. Thus the intent of therapeutic cloning is to heal, not to clone embryos. More specifically, the intent is to clone pluripotent cells from blastomere cells. There is no effort to ensure that these cells are totipotent and therefore capable of implantation for subsequent development into a human clone. Even so, some believe that therapeutic cloning is a slippery slope toward the cloning of humans. Others believe that these therapies are sorely needed to improve the human condition, and that a careful walk along the slope is therefore justified. This difference in beliefs, and in concerns for the embryo, continues to influence the direction of therapeutic cloning research.

Both adult and embryonic cells can be collected and cloned for biomedical research. Adult cells are obtained by tissue biopsy and are then induced to regain the potency they once had as embryonic cells. These cells are clones of the adult from which the biopsy was obtained. Embryonic cells are obtained directly from IVF embryos by blastomere separation or biopsy. Cloning embryonic cells yields cells that are clones of each other; they are not clones of the adults that provided the sperm and eggs. Embryos are normally destroyed in blastomere separation to clone embryonic cells (unless split at about the eight-cell stage to form a cloned embryo or twin), but one technique does avoid destruction of the embryo. Recall that a cell can be removed during an embryo screening procedure called preimplantation genetic diagnosis, and if that cell is cloned, it may then serve both as source for cloning embryonic cells and as a cell for diagnosis, all without apparent harm to the embryo. Thus we can avoid harming the embryo, but can we avoid cloning cells from the embryo? Is it possible to use adult cells, or some other type of cell?

Two techniques described in the stem-cell chapter are important to mention here: nuclear transfer and parthenogenesis. Nuclear transfer is a process where an adult cell nucleus is removed from a somatic cell and then transferred to an enucleated egg. Factors within the egg cytoplasm can then induce reactivation of embryonic genes in the adult cell nucleus to form an embryonic cell (a blasto-mere cell). This type of cell can then be cloned to yield iPS cells. These cells are also clones of each other, but unless the adult cell nucleus and egg come from the same person, these cloned cells are not identical genetic clones of the person providing the nucleus. That is because the enucleated egg still contains cytoplasm with various materials, including extranuclear DNA found in the mito-chondria; these materials are then passed from cell to cell during cell divisions that follow fertilization. Thus iPS cells are genetic hybrids that contain both nuclear DNA of the somatic cell nucleus and mitochondrial DNA from the enucleated egg cell. These hybrid cells are sometimes called hybrid embryos—a term that is mislead-ing. These so-called hybrid embryos are not a result of fertilization; they are, instead, a result of nuclear transfer to form an artificial zy-gote. Rudolf B. Brun, professor of biology at Texas Christian Uni-versity, suggests that this artificial zygote should be called a "somatome," for body part. This is not an embryo; it is an artificially constructed zygote with the genome of a somatic cell nucleus.[11]

To further illustrate the complexity of these issues, and the impor-tance of terminology, consider the fact that the enucleated egg does not have to come from the same species (this is also evidence that related species share important genetic and biochemical processes). For example, Advanced Cell Technology, a private biotechnology company, was the first to report transferring a somatic cell nucleus (in this case from a man's leg) to an enucleated cow egg. The result was a human-cow artificial zygote with the nuclear DNA of a human and the mitochondrial DNA of a cow. The zygote was allowed to develop for only 12 days because, according to a spokesperson for the com-pany, their research is intended to develop stem-cell lines for thera-peutic cloning, not reproductive cloning.[12] Thus, nuclear transfer does avoid the cloning of embryos in the development of pluripotent stem-cell lines, but nuclear transfer does not rule out the possibility of reproductive cloning.

We discussed a second approach to clone pluripotent cells in the last chapter: parthenogenesis. This approach uses laboratory techni-ques to interrupt egg formation so that an egg, which is normally haploid, forms a diploid cell that can then form blastomere cells for the production of cloned pluripotent stem cells. These cloned cells are normally called pPS cells; they are clones of each other and they are clones of the egg donor (albeit not exact, depending on where meiosis is interrupted). Techniques that rely on nuclear transfer or parthenogenesis yield cells that do not use embryos in the cloning process, and thus the terminology that we use to distinguish these cells is important. For precision, cloned hES cells are cloned embryos, cloned iPS cells are somatomes, and cloned pPS cells are pseudozygotes. An iPS cell is an asexual clone of the adult-cell do-nor, and a pPS cell is a clone of the woman providing the egg. These cells "behave" like embryos, and thus the question of whether or not they are in fact embryos is likely to persist. Recall that all cells with a nucleus contain the necessary information for any other type of cell, and even the entire organism, but that information is deacti-vated when it is no longer needed as cells specialize. Technological challenges remain. For example, inducing adult cells to return to an unspecialized state is not unlike the process normal cells go through when they become cancerous—thus it is not clear how iPS cells will function in the body. Nevertheless, the tremendous potential for cell-based therapies while avoiding harm to embryos is a compelling reason to continue research in these areas. Are the reasons for clon-ing entire organisms as compelling?

Reproductive Cloning

The cloning of entire organisms is frequently called reproductive cloning. In nature, the formation of clones is sometimes a reproduc-tive strategy. In the human laboratory, cloning has become a reality. The definition of reproductive cloning to produce children is, according to the President's Council on Bioethics, the "production of a cloned human embryo, formed for the (proximate) purpose of initiating a pregnancy, with the (ultimate) goal of producing a child who will be genetically virtually identical to a currently existing or previously existing individual."[13] Why consider reproductive cloning for humans? It is already possible to transplant tissues and organs without danger of rejection when donor and recipient are well-matched, and antirejection medications are becoming increasingly effective. But if the donor and recipient are genetic clones, then the risk of rejection is nil. Recall the introductory case. Jimmy is several years younger than Allen, and although they are twins, it is apparent that Jimmy desires to have his own identity. But one thing is for sure, Jimmy is the perfect organ donor for Allen—a compelling reason for cloning. This is because clones share allelic forms of genes responsible for their cell-surface markers (or antigens), and therefore when they share transplanted cells, tissues, and organs, their immune systems do not attack the introduced tissue because it is recognized as "self." Joe and Kate cloned Jimmy to save Allen— these children are sometimes called "savior siblings." The technique employed by doctors to clone Jimmy was embryo splitting (or embryo twinning). Blastomere cells can be separated from each other for subsequent development of normal embryonic stem-cell lines, but in Jimmy's case, the cells were then cultured to develop into blastocysts for subsequent implantation and development. Jimmy, a clone or twin, was the result. Someday it may be very efficient in vitro to separate human blastomere cells for subsequent blastocyst development; this would then make multiple cloning (or twinning) of siblings an affordable procedure during IVF. Research has demonstrated success rates as high as 75 percent when mouse blastomere cells were separated and then cultured to form new blastocysts; a second splitting yields similar results, but the potential for success decreases significantly upon the third split.[14] Embryo splitting and nuclear transfer are well-developed technologies for the cloning of animals.

Dolly was not the first animal to be cloned in the laboratory; that distinction goes to a tadpole that was cloned in 1952.[15] But Dolly was a wake-up call for those concerned about prospects for human cloning. SCNT technology is incredibly complex and inefficient; it took 277 attempts at SCNT to yield Dolly. The inefficiency of implantation is a challenge to be overcome in the development of human reproductive cloning. Researchers at the Center for Regenerative Biology and the Department of Animal Science at the University of Connecticut report that, although SCNT does facilitate the development of embryonic stem-cell lines, aberrant reprogramming of the embryo's trophectoderm may preclude implantation for reproductive cloning in humans. They also report many pathologies associated with animal clones that were produced by SCNT pathologies, such as large offspring syndrome, fetal-placental edema, organs of abnormal size, perinatal death, and respiratory problems.[16] Animals continue to be cloned despite these risks. Goats, mice, pigs, and cats are routinely cloned for biomedical and agricultural research, and, believe it or not, it has been reported that a biotech company was planning to hold an on-line auction, starting at $100,000, to clone a pet dog for the highest bidder.[17] Aside from the obvious triviality of such a venture, the first dog, named Snuppy, was cloned in 2005.[18] But will the cloning of pets achieve what pet owners may hope for? Will the cloning of a dying Spot bring back a dog with the same personality? Maybe the same spots, but more than likely, not a dog with the same personality. Spot-2, the clone, would start out as a puppy, and we cannot be sure to what degree a dog's behaviors are learned. And it would require a huge technological leap to accelerate a clone's growth rate so that the original Spot and Spot-2 could stand side by side as look-alikes! Nonetheless, cloning is an efficient way to produce genetically modified animals for research, which can serve as "bioreactors" (living chemical factories) for the synthesis of pharmaceuticals and the growth of human organs.[19] It is possible to transfer human genes to other organisms, and because the genetic code is a universal language, those genes can then be read by the biochemical systems of the particular organism to produce human gene products. For example, erythropoietin is a naturally occurring human protein that is not easily produced in the laboratory or found in nature. Erythropoietin can be used to treat anemia following kidney dialysis, and with plans to increase its availability, rabbits have been genetically engineered to produce human erythropoietin. The term transgenic is used to describe organisms with genes added to their genome from another species. Hundreds of transgenic animals have already been produced. To mention a few: Transgenic cows can produce lactoferrin (an antimicrobial component of natural milk) to enhance human infant formula, transgenic pigs can produce human hemoglobin, and transgenic sheep can produce alpha-1 antitrypsin to treat hereditary emphysema.[20] Pigs offer a source of organs similar to ours, and to increase tissue compatibility, scientists are experimenting with ways to "knock out" (or deactivate) genes that mark the pig's cells for rejection by our immune system.[21] The transfer of cells, tissues, and organs to another species is called xenotransplantation. This is a viable alternative to therapeutic and reproductive cloning in humans, but the possibility of a virus that is specific for one species jumping to another species cannot be ruled out. Another concern is that genetically modified organisms may escape the laboratory setting and introduce new genes or alleles into the gene pool of a natural species. Thus, standards for research protocols to insure safety must remain a priority.

REPRODUCTIVE CLONING: ETHICAL CONCERNS AND LEGISLATIVE LIMITS

Ethical Concerns

The history of reproductively cloning animals is a very recent phe-nomenon. In 1952 a tadpole was cloned, but it was the advent of Dolly, the cloned sheep, in 1996 (announced in February 1997) that shook the world.[22] Not long after Dolly was cloned, several other mammals were cloned: dogs, goats, cats, and horses, to name a few. But most significant, we were now one step considerably closer to the possibility of creating human clones. This new potential raised serious concerns about whether we were playing God, since we were now in a position where we could create life ourselves without the usual means of sexual reproduction. It is interesting to note, however, that secular ethicists tended to use this phrase more than theological ethicists, since the latter viewed the problems as being primarily ethical rather than religious.[23] That is, religion aside, there are significant concerns about the potential results of this technology. In the next section we consider the major arguments in favor of cloning as well as some potential responses to these arguments. In the subsequent section, we outline the major arguments against cloning, as well as some potential responses to those arguments. It is important to note, though, that because of the general opposition to human reproductive cloning, eth-ical and theological arguments both for and against have been given scant attention relative to other genetic technologies. We consider some of the possible arguments that have been made or could be made.

Although we are not yet trying to clone humans, numerous argu-ments support cloning in general. We look at several arguments: the cloning of animals, infertility issues, assisting nontraditional fami-lies, and the autonomy of science as well as that of individuals, and the replacement of organs with little chance of rejection. Those committed to the protection of animals and in favor of animal rights have concerns about our use of animals in this regard. But many also view the cloning of animals as a benefit. Most obvious is the possibility of cloning endangered species. Although there will be issues regarding the lack of biodiversity, at least it would enable us to keep species from going extinct. In addition, some biotechnology companies offer pet owners the possibility of cloning their domestic pets. This could help mitigate the grief that pet owners often experience after the loss of a beloved pet. The cloning of farm animals could ensure that we create individuals with the characteristics we desire. The cloning of research animals may increase efficiency for production of genetically modified organisms. Responses to these arguments include whether we have the right to experiment on animals in the first place; that cloning endangered species will not bring about the result that we would like to achieve of re-creating a robust population of animals that can reproduce themselves as is done in nature; and that there are so many unwanted pets that must be killed that it is a waste of money to create clones of deceased pets, especially because the cloned pet will have the genetic make-up of the parent but not necessarily the characteristics.

An argument that is often presented for cloning is that it would provide an additional option for those with serious infertility problems. Infertility is quite common, especially as women advance in their child-bearing years, and thus all assisted reproductive technology techniques can quite literally be considered a godsend. But cloning would open up additional possibilities. It has even been suggested as a possible option for parents whose last hope at a child results in the fetus dying, and the parents then having the opportunity to clone the fetus and ensure another child of their own. In fact, parents could theoretically create a clone of a dead child.

Related to this argument is the one that in the move toward non-traditional families, especially homosexual couples, cloning could offer the hope of creating a child that is genetically related to at least one of the parents. It can be challenged however, as to why parents would require an exact genetic copy of themselves. In addition, if other assisted reproductive technologies, including the use of IVF, could assist in a child genetically related to one of the parents, why would a clone of one of the parents be necessary?

Another argument in favor of cloning is related to the issue of autonomy of science and of individuals. Regarding science, autonomy is the idea that because the enterprise of science is to continually discover, science should not be hampered in its ability to do so. If scientists want to explore the scientific possibility of cloning in the laboratory, even just to see if cloning of humans can indeed be done, then they should be allowed to do so. The application of this science in technological developments, and future potential ethical problems, should not restrict their search for knowledge. In other words, if it can be discovered, it should be discovered. With regard to individual autonomy, this is related to a liberal approach to genetic technologies in general which maintains that if individuals want to use this technology and have the financial means to do so, and companies are willing to help them with this, then we should not stand in their way.[24] Responses to these arguments are that just because we are able to do something does not mean it should be done, whether by scientists or nonscientists. Also, we will eventually have to deal with the ethical issues raised, and if we already consider them to be significant enough to warrant serious concern, then why should we even proceed down this path?

A final argument is that we might be able to help children already born through the creation of a perfect genetic match for them, particularly with regard to organs. It does not mean that this second "replacement" child or "savior sibling" would exist only as a means to securing the end of saving the life of the first child. They would have their own identity and be their own person, but they could also literally be a life-saver for a sibling without whose help they might die. Responders to this argument are primarily concerned that the cloned child would simply be viewed as a means to an end, which could result in significant social and psychological damage for the child. This response is actually probably the strongest argument against human cloning, so we explore this concern in a bit more detail.

The major concern is with the issue of identity of the cloned child. Let us return to the example for a moment. A major part of the reason that Jimmy was conceived was so that he could help his brother Allen with body parts. This could affect how Jim and Kate view him, how Allen views him, and most important, how Jimmy views himself. Even if Jimmy's parents love him for who he is, it is an inescapable fact that he was partly or even primarily created for the express purpose of being a ready supply of spare parts if Allen should need them. This may have a negative effect on how they would view Jimmy, because they have already demonstrated the potential to put Allen's welfare before Jimmy's. In addition, Allen and Jimmy would not be able to have a normal sibling relationship if Allen had to rely to Jimmy for his health, and if Jimmy knows that at any time his own life or well-being could be compromised for the sake of his brother. Of course the roles could be reversed, where Allen could serve as a donor for Jimmy. But most important, how could Jimmy possibly view himself as an individual, with his own identity, hopes, dreams, and goals? In a sense, Jimmy's destiny was decided by his parents, and Jimmy may have little say in changing his ultimate destiny. Immanuel Kant maintained that an individual should never be used only as a means to someone else's end. One can only imagine what Kant would think of this situation! Of course, there are others who argue that even if Jimmy was a clone created to help his brother, this does not mean that Jimmy is not still his own individual in a real sense. He does have his own hopes, dreams, and desires that may not have anything to do with Allen. In fact, we all come into the world with expectations thrust on us, from which we may not easily be able to escape, if at all. Although Jimmy's situation may be unique, his assistance to Allen's health does not necessarily mean that this will be ultimately detrimental to Jimmy.

A second argument has to do with the commodification of the cloned individual. This is a concern raised with regard to many newer genetic technologies in general. The Catholic *Charter for Health Care Workers* expresses the worry that the resulting child, even of IVF technologies, could be viewed simply as a laboratory product rather than as a unique human being with dignity and created in the image of God.[25] A cloned individual could have the same problem. However, a counterargument could be made that this need not be the case: the genesis of an individual, if they are still human, should not affect the way that we view them or treat them. Consider the birth of the first IVF child, Louise Brown, in 1978. This created quite a stir when it first happened, but now IVF is a very common procedure in which the children born as a result of this technology are not ostracized because of how they were conceived.

A third argument has to do with safety concerns. There are considerable problems with the process of cloning animals, with issues of failed attempts at implantation, fetal abnormalities that result in premature death, and significant health issues in the animals born. Even Dolly required numerous efforts before she was successfully brought into the world. Thus, many think that we should not proceed in the direction of human cloning where significant risks and deformities might await those so conceived. There is not really a strong counterargument to this, except to say that in failed cases of animal cloning, the problem is more likely to occur in the very early stages of development rather than with animals already born. Of course, there would be every reason, on the basis of animal experiments, to proceed very cautiously if and when we decide to go down this road.

Another argument against human cloning is the issue of biodiversity. Cloning does decrease species variability, and species variability is generally considered to be quite important in maintaining healthy and vibrant species populations. Stephen Clark, a Christian ethicist, notes, "Deserts (whether cold or hot) have their own beauty; so indeed does parkland. But it is noticeable that most of us think even gardens better if they are more diverse."[26] Even from an aesthetic point of view, biodiversity is important. But it is also crucial from a biological and environmental perspective. A counterargument to this could be that we would not be using cloning techniques to reproduce the entire species; it would be only in particular cases where there was a compelling reason to clone individuals. Therefore, lack of diversity would not have to be an issue, at least not with regard to humans. Some are uncomfortable about the concept of reproductive cloning of humans. For example, if parthenogenesis is the preferred technique, then this would leave men out of the reproductive picture. If men, whose role has traditionally been essential to the perpetuation of the species, are no longer needed, then what impact will this have on society as a whole in general, for gender relationships, for parent-child relationships, and for men in particular? It has also been argued that cloning is a more efficient method of reproduction than the traditional means. A counterargument to both of these points, though, could be that since sexual activity is one enjoyed immensely by virtually all who participate in it, it is not likely to become obsolete! Finally, a few religious arguments against cloning have to do with whether or not we are interfering with God's role as creator, and whether or not it would interfere with the process of ensoulment. One could argue, however, that humans have been given the task of cocreation with God, and that if the soul exists, it is not necessary that it enter the body only through traditional sexual reproduction.

With regard to the kind of ethical theories that one could use to make arguments, all of the following three are potentially important. One could argue against human cloning on the basis of virtue theories by exploring the intentions and motivations of scientists as well as those who develop and market the resulting technology. Deontologists could argue that there is something intrinsically wrong with human cloning. Consequentialists on either side of the issue would point to the possible beneficial and negative consequences of this practice.

But no matter what one ultimately thinks about the cloning of humans, it is important to remember that genes are not the only things that constitute who we are; environment plays a big part as well. Thus, even natural identical twins who share the same exact genetic material are different in their personalities and are each still considered a unique person. The actual reality of clones, whether artificially created or naturally resulting, would likely be quite different from the science fiction portrayals. Given the fact, though, that there are not many strong arguments in favor of human cloning, as well as the very real dangers with the science and technology at this point even with regard to animals, it is probably not going to be something that we will have to worry too much about, at least not in the immediate future. But we should not let our guard down because today's possibility may become tomorrow's reality, and we need to be ready.

Legislative Issues

It is possible that when it comes to cloning in general, scientific limitations can be addressed more easily than the ethical concerns and statutory limitations. When cloning is involved, a doppelganger comes to mind. Popular culture addresses this issue. *The Island* is a recent science fiction film where so-called "sponsors" pay to support clones of themselves as a future source for tissues and organs. The clones live in a world unaware of their sponsors and under the illusion that some day they would win the lottery for travel to a beautiful "island," where unbeknownst to them, their organs would be harvested.[27] Given these perceptions, and the fact that significant ethical concerns remain when it comes to cloning, it should be no surprise that statutory limitations are likely to play a greater role than scientific limitations in the development of these technologies. According to the Center for Genetics and Society, there are no federal mandates in the United States to prohibit the cloning of humans.[28] However, the National Conference of State Legislatures reports that at least 15 states have passed legislation to prohibit cloning. Specifically, 13 states prohibit reproductive cloning and 6 states prohibit therapeutic cloning research.[29] It appears that in the United States there is a growing effort to ban reproductive cloning, but the mandate is not so clear for therapeutic cloning. Efforts by the U.S. Congress to prohibit cloning are ongoing.[30] However, cloning and embryonic stem-cell research are related issues, and federal funding for embryonic stem-cell research was restricted from August 9, 2001, until March 9, 2009, when President Obama lifted the ban. It remains to be seen how this will influence the direction of research in the United States but Britain's approach to these issues has been different. They established the Human Fertilisation and Embryology Authority (HFEA), with both scientists and lay people in near equal numbers. HFEA has statutory authority and has developed enforceable guidelines that are proving to be very useful as new technologies arise. Ruth Deech, an independent member of the House of Lords, reported that the British government set out to avoid the word "clone" when Parliament was persuaded by HFEA and other organizations to allow nuclear transfer research. Deech suggests that this action enabled biomedical researchers to move beyond embryo-derived stem-cells as the only option for saving lives.[31] More recently, California has taken a similar approach by establishing the California Institute for Regenerative Medicine (CIRM) to provide funds for stem-cell research.[32] It remains to be seen if this type of initiative will have the same kind of influence as HFEA, but we can be certain that scientists are paying attention to concerns regarding prospects for therapeutic and reproductive cloning. The American Association for the Advancement of Science (AAAS) endorses a legislative ban on reproductive cloning. The AAAS cites animal studies as indicative of great risks, and they call for more public dialogue with all stakeholders as the technology advances. The AAAS does support stem-cell research using therapeutic cloning techniques; they cite the potential for tremendous health benefits as a reason, but they also call for greater oversight and a public dialogue regarding the ethical implications.[33]

CONCLUSION

Is the possibility of cloning as troubling as its portrayal in science fiction? Certainly natural cloning through twinning that occurs in humans is noncontroversial as is asexual reproduction in plants and animals. But this does not resolve the issue of how far we should proceed with human cloning in the laboratory. Serious ethical questions remain. Some believe that there are significant benefits of cloning, certainly for animals such as endangered species and even in agriculture, and for those with serious fertility issues who want their own child or possibly even to have a genetic match for another child, as long as the second child is not conceived only for this purpose. But many counterarguments have been offered as well, including problems that already exist with the cloning of animals, the possible commodification of children, and the problem of identity for the cloned individual.

The virtually universal opposition to human cloning is demonstrated both in the strong arguments against the development of its technologies and the virtual ban that exists on such research in the United States. Simply because most states oppose the development of reproductive cloning, though, does not mean that it will not happen in laboratories in other countries. Does cloning of adult humans truly represent a line that we do not want to cross? Or will we eventually be tempted to move down the slippery slope in this direction? Does the world of *Herland* attract us or repel us? One thing is for sure; it is certainly a good use of our time to engage in ethical reflection on this issue, so that ethics will not have to play "catch-up" with science and technology, as is often the case. Joe and Kate are fictitious individuals, but in the future their situation may become a real possibility. From an ethical and theological perspective, however, we must always keep in mind how these technologies will affect the cloned Jimmys of the world.

NOTES

1. Merriam-Webster On-Line Dictionary, 2005, Merriam-Webster, Inc., http://www.merriam-webster.com/dictionary/doppelganger (accessed June 8, 2008).

2. Franklin J. Schaffner, *The Boys from Brazil* (Twentieth Century Fox, 1978).

3. For an interesting recent novel addressing this issue, see Jodi Picoult, *My Sister's Keeper* (New York: Washington Square Press, 2004).

4. For a very helpful and substantive introduction to cloning in general, see Aaron D. Levine, *Cloning: A Beginner's Guide* (Oxford: Oneworld, 2007).

5. Tech Museum of Innovation, "Understanding Genetics," 2004, http://www.thetech.org/genetics/ask.php?id=86 (accessed October 11, 2008).

6. Stephan Reebs, "Time to Split," *Natural History* 117, no. 5 (June 2008): 12.

7. Michael Winterbottom, *Code 46* (Metro Goldwyn Mayer, 2004).

8. Charlotte Perkins Gilman, *Herland* (Mineola, NY: Dover, 1998).

9. Ian Wilmut, A. E. Schnieke, J. McWhir, A. J. Kind, and K. H. Campbell, "Viable Offspring Derived from Fetal and Adult Mammalian Cells," *Nature* 385 (1997): 810–13.

10. *Human Cloning and Human Dignity: The Report of the President's Council on Bioethics* (New York: Public Affairs Books, 2002), 267.

11. Rudolf B. Brun, "Cloning Humans? Current Science, Current Views, and a Perspective from Christianity," *Differentiation* 69 (2002): 184–87.

12. Advanced Cell Technologies is a U.S.-based private company, and they first reported creating a hybrid embryo in 1999 (http://news.bbc.co.uk/2/hi/science/nature/371378.stm). This work sparked debate that continues (http://news.bbc.co.uk/2/hi/health/6251627.stm, accessed October 25, 2008).

13. *The Report of the President's Council on Bioethics*, 267.

14. Karl Illmensee, Khalied Kaskar, and Panayiotis M. Zavos, "In Vitro Blastocyst Development from Serially Split Mouse Embryos and Future Implications for Human-Assisted Reproductive Technologies," *Fertility and Sterility* 86 (October 2006): 1112–20.

15. "Human Genome Project: Cloning Fact Sheet," September 19, 2008, U.S. Department of Energy Office of Science, Office of Biological and Environmental Research, http://www.ornl.gov/sci/techresources/Human_Genome/elsi/cloning.shtml (accessed October 13, 2008).

16. Xiangzhong Yang, Sadie L. Smith, X. Cindy Tian, Harris A. Lewin, Jean-Paul Renard, and Teruhiko Wakayama, "Nuclear Reprogramming of Cloned Embryos and Its Implications for Therapeutic Cloning," *Nature Genetics* 39, no. 3 (March 2007): 295–302.

17. James Barron, "Biotech Company to Auction Chances to Clone a Dog," *New York Times*, May 21, 2008.

18. Levine, *Cloning*, 69.

19. J. L. Edwards, F. N. Schrick, M. D. McCracken, S. R. van Amstel, F. M. Hopkins, and C. J. Davis, "Cloning Adult Farm Animals: A Review of the Possibilities and Problems Associated with Somatic Cell Nuclear Transfer," *American Journal of Reproductive Immunology* 50 (2003): 113–23.

20. Ricki Lewis, *Human Genetics: Concepts and Applications*, 8th ed. (New York: McGraw-Hill, 2008), 386.

21. "Human Genome Project: Cloning Fact Sheet," 11.

22. For the complete story on Dolly, scientifically as well as the social impact, see the book by Dolly's creator, Ian Wilmut (with Roger Highfield), *After Dolly: The Uses and Misuses of Human Cloning* (New York and London: W. W. Norton, 2006).

23. Audrey Chapman, *Unprecedented Choices: Religious Ethics at the Frontiers of Genetic Science*, Theology and Sciences (Minneapolis: Fortress Press, 1999), 93.

24. Two good volumes that promote this kind of liberal approach to genetic technologies are Nicholas Agar, *Liberal Eugenics: In Defence of Human Enhancement* (Malden, MA, and Oxford: Blackwell Publishing, 2004), and Ronald M. Green, *Babies by Design: The Ethics of Genetic Choice* (New Haven, CT, and London: Yale University Press, 2007).

25. Pontifical Council for Pastoral Assistance, *Charter for Health Care Workers* (Boston: St. Paul Books & Media, 1994), par. 24.

26. Stephen Clark, *Biology and Christian Ethics* (Cambridge: Cambridge University Press, 2000), 238.

27. Michael Bay, *The Island* (DreamWorks LLC and Warner Brothers Entertainment, Inc., 2005).

28. "U.S. Federal Policies," 2008, Center for Genetics and Society, http://www.geneticsandsociety.org/article.php?id=305 (accessed October 13, 2008).

29. "State Human Cloning Laws," January 2008, National Conference of State Legislatures, http://www.ncsl.org/programs/health/Genetics/rt-shcl. htm (accessed October 13, 2008).

30. To monitor ongoing efforts in this area, visit www.govtrack.us; for a specific example, see http://www.govtrack.us/congress/bill.xpd?bill=s110-812&tab=summary (accessed October 13, 2008).

31. Ruth Deech, "30 Years: From IVF to Stem Cells," *Nature* 454, no. 7202 (July 17, 2008): 280–81.

32. "California Institute for Regenerative Medicine," 2007, California State Government, http://www.cirm.ca.gov (accessed October 18, 2008).

33. American Association for the Advancement of Science, "Statement on Human Cloning," February 14, 2002, http://www.aaas.org/news/releases/2002/Cloning.shtml (accessed October 19, 2008).

7

My Genes Made Me Do It: The Relationship between Genes and Behavior

"Guilty as charged!" The Judge's gavel sounded as Jeff's heart sank. "But it was not my fault, my genes made me do it! We must appeal." Later that night in his cell, Jeff could not rid his mind of the events surrounding that fateful afternoon. Jeff had a tough childhood, and his behavior often got him into trouble at school. Yet his impulsiveness proved entertaining among friends, and may have even contributed to his success as a rough-and-tumble wrestling coach. Jeff was quick to engage his team in what he called "skull-crushing and tough-love fighting." But this time, in the heat of combat, Jeff lost control and delivered a fatal blow to a student-wrestler. A school security camera located in the gymnasium captured the entire sequence of events, from the time Jeff jumped up from the mat after being pummeled by Bubba, to the point where Bubba's mocking laugh triggered Jeff's sudden and violent outburst. Jeff's defense attorney argued that Jeff did not intend to harm Bubba, but that something very primal within Jeff made him do it. The defense used genetic tests to demonstrate that Jeff carried a less active version of the monoamine-oxidase A (MAOA) gene on the X-chromosome. It has been demonstrated that this less active version of the gene is associated with violent, criminal, or impulsive behaviors in males who were maltreated as children. The defense then provided documentation that Jeff's father was abusive. Thus, Jeff clearly fit the profile for a male predisposed to violent and impulsive behaviors. Was it Jeff's nature that led to his behavior? The prosecuting attorney countered by stating that while variations in the MAOA gene have been associated with violent outbursts, this is only a predisposition, not a certainty. Furthermore, an association, or even a very high statistical correlation, is not causation. How can we be sure that mutations in the MAOA gene cause violent behavior when we cannot even

demonstrate a biological mechanism to produce the behavior? The jury is still out on Jeff's appeal.

INTRODUCTION

In recent years a concerted effort has been made to find the genes associated with certain human diseases; in some cases we know the exact gene that causes a particular disease, and in other cases we know the grouping of genes that can predispose one to particular illnesses. But along with the tendency to try to isolate genes for diseases for medical reasons has been a corresponding interest in trying to find the genes for certain behaviors such as alcoholism and violent behavior as well as a number of psychological disorders such as depression and obsessive-compulsive disorder, to name just a few. This is certainly more controversial than trying to isolate genes for diseases. But one need only go to the OMIM Web site and type in some kind of behavior or psychological problem or human tendency, and you will be amazed about the kinds of behavior for which scientists believe there is a genetic component.[1] Even if we do find genes that are associated with these "disorders," though, this does not necessarily mean that they are only genetically determined, so we must also consider how the environment interacts with genes to influence their expression or nonexpression.

It is widely recognized that the interplay between genetics and the environment is difficult to establish, especially for complex traits.[2] If we are to find genetic predispositions or causes for human behaviors, it raises significant questions with regard to the fundamental component of human nature from the perspective of Christianity. Do we truly have free will or are we actually limited in our choices by our genes? What other implications might a genetic basis for behavior have?

Let us take the hypothetical example of the wrestler Jeff. The reason for Jeff's fatal outburst is what is at question here. As the example indicates, the reality is that there may be a genetic basis for violent behavior, but there are environmental influences and triggers as well. So, if Jeff is primarily influenced or determined by his genetic markers, in addition to his abusive upbringing, can Jeff really be found "guilty?" If his genes played a significant part in making him do this, then how responsible can Jeff really be for his choices? Thus, in addition to challenging the notion of free will, the search for genetic bases for criminal behavior challenges the notion of moral and even criminal responsibility.

Many additional questions can be raised with regard to the possibility that we are partly, largely, or solely determined by our genes. The extreme side of attributing genes for behavior is called genetic determinism—the idea that our genes truly determine who we are and what we do. Thus our genes might be blamed for our bad decisions. Of course, the corollary might be that our genes can be implicated in our good behavior as well. If we are "bad to the bone," as in the refrain of the song made popular by Al Bundy in the television show, *Married with Children*, is it really our fault? What is the relationship between nature and nurture, genes and behavior? What are the implications for morality if we are largely or even partly determined by our genes? What happens to the Christian concepts of morality, sin, virtue, and freedom? If we do not truly have free will, what does this mean in terms of how we compare to the rest of the animal kingdom? Might we be in need of modification of the concept of being created in the image of God?

In this chapter we first explore a scientific perspective on nature and nurture, and then describe the scientific methods in which genes are linked with behavior. We next explore the implications of the foregoing from philosophical and theological perspectives, focusing on the notions of human nature, free will, and morality. We also explore the search for and theological implications of the existence of a "God-gene."

SCIENTIFIC PERSPECTIVE ON NATURE AND NURTURE

What do we mean by nature or nurture in terms of our behavior? Is a nature "versus" nurture approach the best way to examine behaviors, or should we be seeking to understand the nature/nurture dynamic as nature "and" nurture? To address these questions we define nature and nurture from a scientific point of view, explore instinctual animal behavior in terms of learning and sociobiology, and seek common ground with a more holistic view to explain human behavior in terms of nature and nurture.

Scientists seek material causes to explain the influence of both nature and nurture on our behaviors because we operate under the assumption that the natural world is consistently organized according to natural laws. A primary assumption for most biologists is that humans are animals (albeit more sophisticated, supposedly), and that life on earth, including animal behaviors, can be understood in terms of evolutionary theory. Thus biologists attempt to explain nature and nurture as a dynamic relationship that can be understood in terms of our genes, our anatomy and physiology, the environment, and our evolutionary past. Many biologists agree that humans are decidedly different than other animals, but whether or not that difference can be explained in purely physical terms is likely to remain an open question. First, let us define "nature." In *The Splendid Feast of Reason*, S. Jonathan Singer defines behaviors as mentally directed activities by organisms in their interactions with the environment. When it comes to the human organism, Singer suggests that efforts to understand human behaviors have traditionally been the province of psychology, anthropology, and the literary imagination, that is, until recently. Modern biology has demonstrated that elements of behavior are programmed in our genes.[3] Thus, biologists generally agree that nature in the nature/nurture dynamic is defined as the extent to which genes can be shown to influence a particular behavior. We should then be able to demonstrate how those genes function in their environment, both internal and external to the organism, to influence behavior. Note that scientists seek to demonstrate both form and function. Thus, to fully understand the nature/nurture dynamic in scientific terms, we must also be able to demonstrate (define and measure) the nurture part of this dynamic and then explain how it functions in its interactions with genes to influence behavior.

We begin with a simple definition. Merriam-Webster defines nurture as "training, upbringing; something that nourishes, food; or the sum of the environmental factors influencing the behavior and traits expressed by an organism."[4] Biologists expand this definition of nurture to include what they call epigenetic factors that cause changes in gene expression without altering the DNA sequence. Epigenetic factors are particularly important in the development of mammalian brains throughout life. Growing evidence suggests that social and environmental factors may induce epigenetic changes, and that those changes may alter phenotypes during the lifetime of the organism as short-term and long-term somatic changes that are not heritable, and in some cases future generations as germ-line changes that are heritable.[5] Thus, nurture may alter gene expression in our lifetime, and possibly, even future generations. The genes that you received at birth from your parents are like pages in a book of directions, but your future is still open in many respects—you can make choices. Collectively, these genes have all of the information to develop a story of your life, but not necessarily the same story you are experiencing today. Imagine that as time progresses in your lifetime, someone (a social environmental factor) or something (a physical environmental factor) selectively marks pages (genes) in your book for reading—perhaps a teacher reinforces one of your talents. Or as in our example, when Coach Jeff was maltreated as a child, this may have triggered expression of his innate predisposition (possibly the MAOA gene) for violent behavior.

Biologists understand some of the basic mechanisms in nature that control the selective reading of pages in the genetic book of one's life, such as mechanisms that alter how tightly DNA is associated with proteins in chromosomes, thus altering gene expression. Genes vary in the extent to which they are expressed, and it appears that in the nature/nurture dynamic, it is not all nature. The famous evolutionary biologist, Richard Dawkins, is often typecast as a strong genetic determinist, but in his words, when it comes to some behaviors, "Genes have no monopoly on determinism."[6] Philosopher Avrum Stroll considers the question, *Did My Genes Make Me Do It?*, in his recent book on philosophical dilemmas. He concludes that this is an unanswerable question in light of free will.[7] For scientists, the philosophical concept of free will does not easily lend itself to empirical test without our knowing more fully the role that our genes play in determining behavior. This lack of knowledge may cause us to perceive a tension between our inborn constraints and our ability to be free of those constraints—our nature versus our nurture.

The nature versus nurture approach might not be the best way to seek an understanding of the nature/nurture dynamic. When we consider nature in opposition to nurture, we are encouraged to take an either-or position. But what about the possibility that this is a false dichotomy and that the relationship is best described as nature "and" nurture? The question should be to what extent do factors internal to the organisms (its inborn form and function) and factors external to the organism (its living and nonliving environment) interact to influence behavior? Consider what toads and humans have in common as an example of an inborn behavior. The German neurobiologist, Jörg-Peter Ewert, studied the neurological circuitry of European toads and demonstrated that there was an ideal stimulus for triggering an innate prey-capture behavior (an inborn form and function for food-getting). Using bar-shaped images to simulate moving worms and crickets in a laboratory setting, Ewert observed that these toads would only extend their tongue in an attempt to capture a bar-shaped image that was about 4 to 16 times longer than wide and traveled in a very specific motion; any other shape or motion caused the toad to flee. Thus, a very specific stimulus triggered a sequence of behaviors: even when a toad retracted its empty tongue, the toad would close its eyes while swallowing and wipe its eyes with one hand.[8] This innate species-specific action is programmed in the genes—prey-capture is a fundamental animal behavior. But what does this have to do with humans? Consider the fact that we all have a little bit of toad brain in each of us because the vertebrate brain evolved through a series of add-ons with our brain stem as an ancient source for our reflexive actions such as breathing and swallowing. Yet we do not reflexively reach out in an attempt to capture food of a given size and shape, although grabbing a piece of rich chocolate is a possibility for some of us. But one thing is for sure: if we fail to capture that bar of chocolate, we do not close our eyes and swallow while wiping our mouths with a napkin. And if we do capture that bar of chocolate and begin to swallow, once the candy is in the back of our throat and about to go down, we can no longer control the swallowing process. Is this what is left of our prey-catching instinct? If we are to understand human behavior, then we must take into account the fact that, whether we like it or not, the form and function of our body are of animal origins. The process of evolution is not quick to discard a functional complex structure such as a working brain stem; instead, we have benefitted from newly added layers of brain tissue during the evolution of fish, amphibians, reptiles, birds, and then mammals, and ultimately our species. This extra brain tissue facilitates learning in addition to automatic responses to environmental cues.

It would be difficult to teach a toad to respond differently, but so-called higher animals do have the added brain capacity to learn a behavior. The work of Ivan Pavlov and B. F. Skinner are mentioned here to illustrate this point. Pavlov was a famous Russian physiologist studying mechanisms of digestion when he observed that dogs would begin to salivate as soon as they observed the researcher preparing food in advance of actually receiving the food. Pavlov then designed an experiment to see if the dogs could learn to salivate in response to some other stimulus. He then demonstrated that if a bell (the conditioning or training stimulus) was presented just before the food (the unlearned or unconditioned stimulus), eventually the dog would automatically salivate upon hearing the bell, even when the food did not immediately follow the bell. This came to be called classical conditioning. Skinner's work went even further than Pavlov's by demonstrating that learning was not limited to a series of reflexive responses to stimuli. Instead, animals during their trial-and-error activities could learn to associate events with an ultimate reward such the release of food when rats touched a lever in the famous Skinner Box—this type of learning is called operant conditioning.[9] We can easily describe instinctive actions (such as a toad catching prey), and we can illustrate what appears to be innately influenced learned behaviors (as we have seen in dogs and rats), but we are a long way from demonstrating the material causes for self-awareness and thoughts that influence complex human behaviors. Dualists believe that the mind and body are fundamentally different (this is the age-old mind-body debate), but most scientists are materialists and therefore generally operate under the assumption that the mind and body are the same thing. Thus, scientists seek biological causes for behavior.[10] Earlier we mentioned with reference to the human brain that biologists are seeking ways to understand how nurture may alter gene expression in our lifetime, and possibly even future generations.[11] Biologists are also taking into account the possibility that evolution may act on these variations in gene expression in the social context.

Sociobiologists, unlike traditional biologists, seek to understand human behavior by taking into account our biology *and* our society. The fundamental units for analysis include genes, memes, and the system as a whole. The field of sociobiology provides a framework for this analysis. It is generally agreed that E. O. Wilson's Pulitzer Prize-winning book, *On Human Nature*, first published in 1978, marked the beginning of sociobiology as a discipline. The book was controversial when first published because it challenged a view that the human mind is primarily a product of learning and experience. Since that time, Wilson's lifelong study of social insects and his expertise in biology, coupled with a gift for communicating his interdisciplinary insights, enabled him to develop a persuasive argument for a more naturalistic view of the human mind and our nature. (The book was republished in 2004 with a new preface.) Wilson explains that neurobiologists seek to explain "how," and the evolutionary biologists seek to explain "why."[12] As a biologist, Wilson offers an evolutionary explanation as to how our materials-based culture may have developed over time. He proposes that when prehumans began to walk erect and their hands were free to handle objects, then both the gradual development of human intelligence and the ability to manipulate objects could "mutually reinforce" each other to ulti- mately yield the materials-based culture in humans today.[13] The gene is the basic unit for transmission of our genetic predispositions to the next generation. The meme (a term coined by Richard Daw- kins) is the basic unit for transmission of culture.[14] Both the gene and the meme are stable, yet capable of variation, and over time they may interact as fundamental units on which natural selection can act. For example, do epigenetic factors, as material causes that we can measure, influence the expression of our genes and our cul- ture, and if so, how?

Growing evidence suggests that social and environmental factors may induce epigenetic changes in the way our genes are expressed, and that in addition to altering phenotypes during the lifetime of an individual, future generations may also be affected. Studies suggest that stress and drug addiction may serve as epigenetic factors with subsequent changes in learning and memory. These types of changes reflect the brain's plasticity (the ability to reassign functional areas following injury) and serve as evidence of epigenetic changes (par- ticularly in light of the fact that this is dedicated neural tissue). Genomic imprinting is epigenetic and occurs when genes are imprinted, turned on or off, and then passed to the next generation. Males can pass imprinted genes along with their sperm, but the maternal influence is more far-reaching because eggs have the abil- ity to reprogram imprinted genes to restore totipotency (recall that this occurs in SCNT). It is probable that epigenetic mechanisms will be demonstrated as contributing to psychological disorders (such as schizophrenia, depression, OCD, traumatic stress, and addiction), clinical conditions (such as obesity, cancer, and neona- tal programming for cardiovascular disease), and stem-cell reprog- ramming. Thus, focusing on variations in our genes alone will not be adequate; we need to also understand the epigenetic processes.[15] Imprinting in animals is another example of the epigenetic process, specifically, the plasticity of a developing brain. Konrad Lorenz demonstrated that when newly born animals (ducks premiered in his classic study) are raised by another species, they acquire some behaviors of the nurturing species.[16] For example, imagine that you incubated a group of duck eggs until they hatched, and only you were involved in their early days as a duckling. These ducklings would then become socially bonded to you as a parent figure and follow you around as if you were a mother duck. Explaining these kinds of epigenetic processes may require more holistic approaches, a departure from conventional scientific methods. Steven Rose, in *Lifelines: Biology Beyond Determinism*, acknowledges that the reductionist approach is effective for the disciplines of physics, chemistry, and molecular genetics. Mapping the human genome would not have been possible without reducing DNA to its fundamental parts. Rose proposes an alternative framework to understand life's processes. He discounts sociobiology and biological determinism and argues for a perspective that transcends genetic reductionism to view life as organism-centered, not gene-centered. Living systems do not easily lend themselves to causal explanations that are most often explained in terms of a temporal chain of events.[17] For example, the actions of genes located at 13q14-q21, 14q22, and 4p13-p12 are associated with alcohol dependence (OMIM #103780), but the action of these genes has not been placed in a temporal chain of events to cause the dependency; this would be the explanatory mechanism.[18] Since both nature and nurture influence dependency, the biochemical chain of events leading to that dependency is difficult to demonstrate. Is it possible that alcohol consumption results in metabolites (breakdown products) that induce greater pleasure for some people? Does brain chemistry change over time under the influence of alcohol and social interactions? Do the actions of family and peer groups affect our biochemistry? These are not nature *or* nurture questions; they are nature *and* nurture questions. When it comes to human behavior, our genes do not necessarily "make us do it." Thus, biology is not destiny.

FINDING GENES FOR BEHAVIORS

Mapping genes to precise locations on specific chromosomes has ushered in a wave of efforts to associate mapped genes and behaviors. For up-to-date information on a behavioral trait (a phenotype) and efforts to link genes (a genotype) to that trait, simply visit the OMIM Web site and type in the name of the trait.[19] For example, type in the term "homosexuality" and then click on the OMIM #306995 assigned to this gene (Homosexuality 1; HMS 1). A page will load showing that the gene maps to locus Xq28 (a symbol indicating that this gene is on the X-chromosome, "q" or long arm, location number 28).[20] The OMIM page has a thorough annotated bibliography of the current research on this gene. It would be inappropriate to say that this gene causes homosexuality; instead, it can be said that the gene may in some way influence the expression of homosexual orientation. When geneticists seek to understand the influence, if any, that a gene might have on behavior, we call that gene a candidate gene for the particular behavior. This is a work in progress for most types of behavior, and therefore it is necessary for our discussion here to focus on how scientists seek to understand the influence of genes on behavior.

In this section we introduce scientific methods used by behavioral geneticists and describe some current examples of ongoing research attempting to link genes and behavior. Geneticists who study human behavior generally use two approaches that are broadly categorized as quantitative and molecular. Quantitative approaches rely on twin, adoption, family, and population studies to determine if there is an association between an observed behavior and a pattern of inheritance. Molecular approaches use laboratory techniques to identify a specific gene or genes believed to be associated with a particular behavior. Both approaches employ statistical methods in an attempt to determine if there is a significant correlation between a particular behavior and observed biological traits, and specific genes if known. Efforts are then made to understand any interactions between the genetic and environmental factors associated with the behavioral trait. This is a very complex task. Several specific examples will illustrate how difficult it is to determine if a behavior has a genetic basis; that is, to identify a gene and how it works (the nature part), and to then consider to what extent the environment affects its expression (the nurture part). Key concepts such as heritability and quantitative trait loci are defined as they are introduced in the studies of parenting behaviors, smoking behaviors, attention deficit hyperactivity disorder (ADHD), and autism.

Heritability is a term that scientists use to describe the proportion of phenotypic variation within a group that can be attributed to genes (nature). Geneticists use statistical methods and a genetic concept called concordance to estimate heritability. Concordance is the percentage of paired twins exhibiting an observed trait. If genes play a greater role than environment, then we expect to see a higher percentage of identical (monozygotic, or MZ) twins than fraternal (dizygotic, or DZ) twins with a given trait. For example, autism is a complex multifactorial trait because it is believed to be caused by several genes, with their expression somewhat dependent on the environment. This is because concordance for autism is estimated to be 90 percent in MZ twins and 4.5 percent in DZ twins—more nature than nurture. In contrast, concordance for facial acne is estimated to be 14 percent in MZ twins and 14 percent in DZ twins—mostly nurture. Twin studies, including twins who are raised separately, and adoption studies are important tools for estimating heritability. If genes play a greater role than the environment in the expression of a trait, then the heritability score is high. For example, the heritability for height is estimated to be 80 percent genetic and total fingerprint ridge count is 90 percent genetic. If genes play a lesser role than environment in expression of a trait, then the heritability score is lower. For example, the heritability for mathematical aptitude is estimated to be 30 percent, spelling aptitude is 50 percent, and verbal aptitude is 70 percent.[21]

A recent study illustrates how scientists attempt to establish heritability for a complex trait such as ADHD. Two approaches were applied: diagnostic data (measures of observable behaviors) and quantitative trait loci (QTL) analysis. QTL refers to sequences of DNA associated with a trait. These loci are used to identify candidate genes for the trait under investigation. Efforts to identify candidate genes for the ADHD phenotype (hyperactivity-impulsivity) have had moderate success, but gene variants in the dopamine receptor system are candidates. Because ADHD is an extreme form of a wide range of continuously varying behaviors in a population and multiple factors are involved, it is difficult to demonstrate the effects of one or more genetic variants. The researchers in this study set out to consider both environmental and genetic components for hyperactivity-impulsivity with three behavioral measurers (a composite index when combined): motion detector data, a standardized teacher rating scale, and a standardized parent rating scale. Parent and teacher ratings of ADHD behaviors are subjective measures and are therefore less reliable, but recent developments in techniques that use motion detectors to directly measure the activity levels of subjects can produce more objective and reliable measurements. A composite index of the data was then used for comparison of hyperactivity-impulsivity in 325 MZ twins, 253 same-sex DZ twins, and 258 opposite-sex DZ twins, ages 7 to 9. The heritability score for hyperactivity-impulsivity was observed to be 77 percent for the composite index. Thus, composite measures may facilitate more objective measures for the statistical analysis necessary to demonstrate an association between behaviors and candidate genes.[22] Quick and easy ways to measure the genotypes of individuals are also necessary.

Advances in genotyping technologies have facilitated the search for candidate genes associated with behaviors and other traits. A common tool in genotyping is the gene chip (DNA microarray), so named because it reminds one of a computer chip, but with the capacity to probe a sample of DNA for specific sequences. Thus, it is easy to screen one's DNA sample for gene sequences using the appropriate gene chip. Genetic linkage analysis begins with a search for genetic variations (or markers) in the genomes of families with the observed trait. A classic example (before the days of gene chips) is Nancy Wexler's hunt for a genetic marker associated with Huntington disease. Wexler, a clinical psychologist, understood the disease on a professional level, and because her mother and several relatives died of the disease, she also understood the disease on a personal level.[23] She examined the inheritance pattern for this disease in an isolated population of Venezuelan villagers, where a sailor brought the disease to their village many generations ago. Wexler was able to identify a genetic marker on the chromosomes that corresponded to the observed inheritance pattern (today we can test for the gene directly). This approach does not require a priori knowledge of mechanisms for gene expression, but genetic association studies typically begin with some prior knowledge of specific genes that are thought to be involved in the expression of a particular trait (candidate genes). For example, researchers studying smoking behaviors suspect that genetic factors influence smoking persistence and dependence. Thus, they begin by examining tobacco-dependent and nondependent individuals for the prevalence of candidate genes. Many genes are under consideration as factors in smoking dependency, including gene variants in the dopamine and serotonin pathways.[24]

The next example illustrates how researchers attempt to link candidate genes to a behavior. A team of researchers suspected that two dopamine-related genes (COMT, catechol-O-methyltransferase; and DRD4, a dopamine receptor gene) might influence parenting behavior in stressful situations. The genetic component of their study involved genotyping 176 mothers to determine if they were carrying variant forms of the two dopamine-related genes (COMT-val, DRD4-7R). To measure stress and parenting behavior, mothers were rated on a scale for potentially stressful events in their lives, and they were observed assisting their children in problem-solving tasks. Mothers with both variant forms of the dopamine-related genes (the COMT-val and DRD4-7R alleles) were observed to be less sensitive toward their children when dealing with stressful events in their lives.[25] To say that these genes "cause" mothers to be less sensitive would not be appropriate. We can only say that there is a statistical relationship between the candidate genes and the behavior as defined with the particular tools used to measure behavior. Today it is more challenging to precisely measure a behavior than it is to measure the presence or absence of genetic variation. This is because genotyping technologies are well established, but reliable and valid measures have not been established for all behaviors. What is the next step if a gene is linked to a behavior?

When a gene or genes are found to be associated with a particular behavior, such as smoking or parental behaviors, then it is still necessary to identify biological processes (or mechanisms) within the body to demonstrate how these genes are influenced by the environment and expressed over time—not a small task. The following example illustrates this challenge. Autism spectrum disorders (ASD) is characterized by a range of phenotypes, including social, linguistic, and motor disorders. Evidence suggests that genetics, and to some extent environmental factors, play a role, but the mechanisms remain unclear. Researchers have found a statistical association between several alleles for the oxytocin receptor gene (OXTR) and ASD in humans. Oxytocin (OT) is part of the prolactin system (milk production) in mammals, and also acts as a neurotransmitter. Maternal OT during childbirth has been linked to gamma-aminobutyric acid (GABA) activity in the fetal brain, and the observation that GABA signaling problems are associated with ASD suggests an indirect role for OT in ASD. Supporting evidence is the fact that lower levels of OT have been observed in the plasma of autistic children, and some ASD behaviors are reduced when OT levels are restored. Many other candidate genes are also under consideration.[26]

However, linking genes and behavior is only a first step; scientists are finding that it is very difficult to demonstrate how genes function to influence a particular behavior.

Matt Ridley, in *Nature via Nurture: Genes, Experience, and What Makes Us Human,* provides a persuasive argument and evidence that genes are both causes and consequences.[27] According to Ridley, one of the many big surprises upon mapping the human genome is the small number of human genes compared to original estimates, and the vast number of genes we have in common with other animals. It is now obvious that a group of genes called the HOX genes came into existence early in the evolution of animals. These genes prescribe the body plan in fruit flies, and, as it turns out, those same genes along with some others orchestrate development in higher animals as well. So once the genes for a body plan were worked out by nature, slight modifications in expression during development could yield a range of body types. The HOX genes work by producing transcription factors, which when they combined with a region of a gene called the promoter, switch that gene on and off for transcription. Once a complement of genes was in place to develop organisms by turning promoters for structural genes on and off at given times, then natural selection needed only to act on minor variations in the structural genes to which promoters were attached to build very different body plans. We have about 99 percent of our genes in common with the chimpanzee, but the challenge will be for scientists to determine how and when genes are turned on and off during development, and throughout our lifetime, to make us so different from as well as similar to the chimpanzee.[28] Ridley suggests that Jean Piaget's cognitive development model in children is one example of how genes can be both causes and consequences. Piaget believed that mental structures for learning were genetic (nature), and that feedback from experience (nurture) was necessary for normal development.[29] Thus genes may be deterministic in that they can produce consistent products, but the mechanisms for switching genes on and off in response to internal and external environments, including the way we are treated, make them what Ridley calls "mechanisms of experience."[30]

The MAOA gene variant carried by Coach Jeff in the introductory example shows how our genes may be influenced by our experience. First we consider a classic study described by Ridley, where it was determined that 442 boys (born in the early 1970s in the city of

Dunedin, New Zealand) who were maltreated as children were more likely to exhibit violent and antisocial behaviors. Researchers Terrie Moffitt and Avshalom Caspi began by examining the boys for low and active forms of the MAOA gene; the low and active forms are a result of mutations within the gene's promoter region responsible for gene expression. Controlled animal studies provide support for this mechanism; mice with the MAOA gene deactivated exhibit aggressive behavior, but normal behavior is restored when gene activity is made active again. Moffitt and Caspi observed that if men were maltreated as children and had low-active MAOA, they were four times more likely to have committed rapes, robberies, and assaults. In contrast, they observed that if men were maltreated as children and had high-active MAOA, they were not likely to exhibit these criminal behaviors. Thus, it is not genes alone nor is it environment alone that is responsible for this behavior. It is genes and environment; it is nature and nurture.[31] The defense attorney for Coach Jeff argued that because Jeff carried a less active version of the MAOA gene, and he was maltreated as a child, he was prone to violent behavior. As a challenge to the defense, the prosecution argued that a mechanism to explain how this gene influences behavior has not been demonstrated, and therefore it is not possible to say for certain that Jeff's genes caused him to commit murder. What might the ruling be if scientists find a way to demonstrate causality in such a case? Would a case like Jeff's be reconsidered? How would society react to the possibility that someone is not held accountable for certain behaviors? These ethical questions make the science seem easy! It is to these questions that we now turn.

IMPLICATIONS OF A GENETIC BASIS FOR BEHAVIOR

"The fault, dear Brutus, lies not in our stars, but in ourselves." Thus spoke the main character in Shakespeare's play, *Hamlet*. A contemporary corollary has been offered by James Watson: "We used to think that our fate was in the stars. Now we know, in large measure, our fate is in our genes."[32] What they both seem to agree on is that something in our very nature makes us who we are. It is clear from the previous two scientific sections that there are ways to study the influence of genes in behavior. It is also clear that we cannot simply look at nature (genes) but that we must also look at nurture (environment). Assuming, however, that there is a genetic basis for many behaviors, including violent behavior, what might the theological and ethical implications of this be? In this section we look at the concept of human nature, the concept of free will (these two are closely related, especially in theology), and the implications for morality of a genetic basis for behavior, especially for "bad" behavior.

Much has been written on the concept of human nature, both philosophically and theologically.[33] Before we can make assertions about human nature, though, we need to ask how we can study it. The main way is by observing human behavior. Evolutionary psychologists have focused on several areas for inherited tendencies, in particular mental disorders, addictions, intelligence, violent crime, sexual orientation, and personality.[34] Christian theology developed its doctrine of human nature using the sources of Christian ethics described in an earlier chapter: What does the Bible say, what has tradition passed down, what is our experience, and what does our reason contribute to our understanding? In a way, to study human nature is to apply the same kind of study of animals to humans. When we consider the cat, most of us have a basic understanding of what "cat nature" is and how, for example, it differs from "dog nature." Thus, to describe the nature of any species is to look for what is common to all members of a species—that is, the essence of that species—and to determine how it differs from other species. Thus, a key feature of human nature, by almost anyone's understanding, is that we are not governed only by instincts and therefore are not a slave to them in the same way as the toad is.

Not everyone agrees, though, on what the essence or features of human nature are, or to what extent the agreed on features are present. An important component in the discussion of human nature is the concept of what is "normal" for human beings. If something is normal, then we might be able to say that it is part of what it means to be human. But of course that does not allow for sufficient variability. "Normal" can be understood statistically: as the average, the median, or the mean. It can also be understood socially, relating to ordinary notions of what people believe is acceptable. It can also be understood biologically, by tying it into biological function and dysfunction.[35] But how ultimately do Christian and secular perspectives understand human nature?

From a traditional Christian perspective, human beings and hence human nature have been created in the image and likeness of God, unlike any other creatures. Though there has been much speculation with regard to what it means to be created in the image of God, it can be described broadly by separation into two categories: ontological and functional. The ontological approach tries to determine what it is that is part of the being or essence of humans, ultimately given to us by God, and has usually been understood to include the concepts of rationality, spirituality, free will, and moral freedom.[36] The emphasis in the ontological approach is to determine characteristics that separate humans from animals. The functional approach focuses more on the task given to humans (by God) to have "dominion" over the rest of creation rather than with any reference to what the essence of humans is. However, a particular function requires a particular ontology, and thus ontology has traditionally gotten more attention.

Although there is considerable agreement among Christian theologians as to what constitutes human nature, the secular realm is much more divided. Of course, depending on what one thinks is the essence of human nature will affect how malleable one considers it to be. In the book *Seven Theories of Human Nature*, Leslie Stevenson outlines some of the rival theories that have had their proponents: The philosopher Plato viewed humans dualistically, as social creatures, and thus saw education as the primary means of perfecting individuals. Marx believed the alienation that humans experience is rooted in economic conditions rather than in a metaphysical or religious reality. Freud viewed humans as largely captive to their instincts and drives, especially on those of power and sex, although psychoanalysis could be an aid in helping individuals to create harmony among the various parts of the mind. B. F. Skinner thought that humans could be understood only with regard to their behavior and not on the basis of mental entities, and that we have an innate aggression toward our own species. Jean-Paul Sartre believed that the freedom of individuals was the most important human property.[37] And these are just a few possibilities! Needless to say, there are many different answers from a secular perspective regarding what human nature is. But most secular, or philosophical approaches, would agree with the Christian perspective that reason is a key component of what it means to be human.

Implicit in this understanding of reason, though, is also the idea of free will. Free will is both a philosophical and a religious concept, and there is considerable overlap in their approaches to the question. Thus, a key issue is to what extent we are truly free. From both a philosophical and a religious perspective, this is usually discussed with regard to the notion of determinism. Are humans truly free to make their own moral choices, or are they constrained in some way, such as by their environment, or by their genes, or by their God? The philosopher Avrum Stroll nicely outlines three basic philosophical theories with regard to free will: hard determinism (in which no humans or animals are ever free to choose or act, and everything that happens is inevitable and unavoidable), indeterminism (in which all material or physical events are strictly determined by antecedent forces, but the mental domain is free from these kinds of constraints), and soft determinism (which maintains that determinism or causality and predictability are compatible with free will). Many Christian theologians would resonate with the third option because it enables the Christian tradition to hold simultaneously to the idea of God's omniscience (knowing everything) and foreknowledge, as well as with the idea that God still allows humans to choose their actions freely in the moment. Let us take the example of the sin of Adam and Eve. Although God could have known that Adam and Eve would sin, that knowledge still did not infringe on the free choice of Adam and Eve (and subsequently of any humans) to disobey God.[38] In any case, Christian theology rests on the foundation of human moral freedom. Humans are free to choose good, but they sometimes choose evil. In fact, the doctrine of theodicy, which tries to explain why evil exists if God is all-loving and all-powerful, is usually answered by what is called the free-will defense; that is, the evil caused by human actions (e.g., war, rape, theft) is not God's fault, but rather humans who choose to go against what they know to be right. In some ways, the Christian tradition has a much stronger notion of free will than do many of its philosophical counterparts. Of course, original sin is still an important component of Christian theology, which suggests some kind of tendency in the direction of evil, at least at times, but it never completely eliminates the idea that humans still can resist successfully this tendency.

So what does this all have to do with a genetic basis for behavior? Any understanding of human behavior that would somehow affect the ability of humans to make free moral choices would be problematic, or at least force us to reconsider our understanding of Christian anthropology. Traditional Christianity tends to reject what does not fit in with our predetermined, or long-held, theological and ethical beliefs. However, what may be needed at times is for the Christian

tradition to revisit and possibly reformulate some of the basic doctrines or beliefs of our faith in light of new discoveries, including
genetic ones. Let us take the example of Darwin's theory of evolution. Those parts of the Christian tradition that have accepted this
theory as the best explanation of how life arose needed to modify
some of their strongly held convictions as a result, such as conceptions of humans as significantly other than animals and tenaciously
holding to literalist interpretations of the creation account as found
in the book of Genesis. The fact that many Christians still reject
evolution and see it as incompatible with a Christian understanding
of creation demonstrates how much of a stronghold traditional
Christian doctrines have on believers, even in light of evidence to
the contrary. So, if we do find that there is a strong genetic basis for
moral and immoral behavior, for example, we may need to rethink
how "free" we truly are.

What, then, are some ethical implications for discovering a
genetic basis for different behaviors? Obviously, it would affect how
we view others, or even ourselves, if we know there is a genetic basis
for at least some behavior. Let us return for a moment to our example. If it can be scientifically demonstrated that Jeff has a strong
genetic predisposition (as well as strong environmental influences)
that has resulted in him being more likely to commit a violent act
than someone who does not have this genetic predisposition, we
probably should not judge Jeff as harshly. This would certainly affect
our understanding of those judged to be criminals if we believed
that they were more likely to commit acts because of their genetic
make-up.[39] In addition, it would force us to rethink the Christian
idea of sin and moral responsibility. Although Christianity has not
typically judged actions only in light of the actions themselves, and
has also considered important the character of the agent as well as
the consequences of the action, it would modify our tendency to
largely judge at least some actions on the basis of their intrinsic
wrongness. If there truly were a genetic basis for bad behaviors, then
it seems that Christianity could also (or instead of modifying its
ideas of right and wrong actions) support genetic therapies to help
individuals with these tendencies or predispositions. Since healing is
an essential part of the Christian tradition, with Jesus as the role
model, any attempt to heal humans even of genetic maladies that
have the potential of harming other humans seems like a good idea.
Overall, though, discovering a genetic basis for behaviors should

make us less judgmental with regard to anything previously considered problematic in the Christian tradition. A genetic basis for alcoholism at least partially confirms the disease theory rather than viewing it as an immoral action. A genetic basis for homosexuality might help us to see this sexual orientation as part of the normal variation of human sexual behavior, and might enable us to be more accepting of those who have previously experienced much judgment from the Christian community. Of course, it could also lead to the opposite situation: that some would want to screen out or modify this tendency. But who decides? We are then back to the issue of designer babies and whether we should be doing any screening at all. Thus, although a genetic basis for some behaviors would challenge some of our long-held philosophical and theological ideas about human nature and free will, we should not fear greater knowledge about who we are, even if it comes from the hands of scientists rather than theologians!

IS THERE A SO-CALLED "GOD GENE"?

We are conflating the disciplines of biology and theology when we dare to ask such a question! Rephrasing the question may allow a biologist and a theologian to remain true to their respective disciplines. Is it possible that there is a genetic predisposition for behaviors that we might describe as spiritual? The example that follows illustrates the challenges that scientists face as they attempt to identify candidate genes for a particular trait, and then set out to identify the mechanism. The concept of a genetic basis for spirituality is certainly not unique to the biologist Dean Hamer. We thought it would be interesting to explore his endeavors, which he wrote about in a book that created quite a buzz in many circles when it first came out. We consider first the way that this study was scientifically carried out, and then consider some theological issues.

In his famous book, *The God Gene: How Faith Is Hardwired into Our Genes*, Dean Hamer sets out to demonstrate a material association between a single candidate gene and personality traits believed to be associated with spirituality.[40] His story serves as an excellent example of how scientists work to understand genes and behavior. Hamer set out to demonstrate a material association between a single gene (acknowledging that there may be others) and personality traits believed to be associated with spirituality. He begins with two assumptions. First, genes predispose one to spirituality, but religion is perpetuated by culture (a meme). Second, spirituality is based in our consciousness (where we can seek a genetic component), but culture is based on cognition.[41] To demonstrate a genetic basis for spirituality, Hamer needed to find a research sample and a valid way to measure spirituality so that he could identify candidate genes. Hamer and a team of scientists at the National Institutes of Health were examining the genetics of smoking behavior and associated personality traits for a study that was sponsored by the American Cancer Institute. Each of the participants in the smoking behavior study had provided a DNA sample and they were given the valid and reliable Temperament and Character Inventory (TCI), developed by Robert Cloninger for his research on the origins of personality. A portion of the TCI includes questions to gauge self-transcendence across three areas: self-forgetfulness, transpersonal identification, and mysticism. Hamer realized that spiritual feelings are often described in transcendent terms and that participant TCI responses could therefore serve as a measurement for spirituality.[42] David Commings, a psychiatric geneticist, had already reported several candidate genes for TCI traits, including self-transcendence; his work, and other studies on mind-altering drugs, caused Hamer to focus on monoamines. Monoamines are neurotransmitters such as dopamine and serotonin. In consultation with George Uhl, a scientist at the National Institute of Drug Abuse, Hamer turned his attention to several variants in the VMAT2 gene that is responsible for packaging a host of monoamines for signaling in the nervous system; subsequent analysis found an association between the VMAT2 variants and self-transcendence.[43] The next step was to propose a mechanism for this gene's expression.

Hamer used Nobel Prize winner Gerald Edelman's empirically based theory of consciousness to propose a mechanism for the VMAT2 gene. According to Edelman's theory, communication between two key components of the brain, the thalamocortical and limbic brain, are responsible for consciousness. Evolution of the brain linked these systems in such a way as to promote higher levels of consciousness in humans. The older limbic brain, in evolutionary terms, senses the environment and then sends signals of value to the newer thalamocortical region of the brain. The thalamocortical region then values and categorizes that information for feedback to the limbic-brain system for further categorization. The results of this feedback system yield what Edelman calls a "remembered present," or consciousness. The VMAT2 gene packages several different neurotransmitters that are responsible for linking the limbic and thalamocortical regions. Hamer's data demonstrated a statistical link between the VMAT2 gene and personality traits associated with the interactions in the limbic (for primary consciousness) and thalamocortical (for higher consciousness) regions. The next step was to demonstrate that VMAT2 is essential and does in fact play a role in the release of serotonin, dopamine, and noradrenalin. For this purpose, it was necessary to bioengineer "knock-out" strains of mice that were homozygous and heterozygous for a nonfunctional (or "turned-off") version of the VMAT2 gene. These mice were inactive, uninterested in food, and frequently died prematurely, and because all of the mice had very low levels of serotonin, dopamine, and noradrenalin, the physiological role of VMAT2 was confirmed.[44] This well-constructed study by Hamer and his associates exemplifies the type of work necessary to identify genetic predispositions and mechanisms for gene expression. That said, this work provides no indication as to whether or not there is a God, and clearly that was not the intent of his study, but it does provide some insight into the possibility that some of us may be more predisposed to spiritual experiences (as measured on the TCI) than others. But what might the implications be for theology?

Before offering some comments on his conclusions, it is important to understand his argument. In his book, he distinguishes spirituality from religion, avoids reductionism by asserting numerous times that there is not one gene for spirituality, and clarifies that he is not trying to prove whether God exists, but rather to account for why people believe. He defines spirituality as having three characteristics: transpersonal identification, self-forgetfulness, and mysticism. Although spirituality as defined can be attributed to certain genes, meaning that spirituality is genetic, he maintains that religion is cultural (mimetic; recall Richard Dawkins' understanding of memes). Hamer came upon the concept of a spirituality gene as he was studying cigarette smoking. He maintains that all humans (but not animals) have this predisposition for spirituality, but some more than others. Spirituality can be compared to the trait of height; it is not an either/or, but a spectrum along which people fall. In addition to how strong a particular individual's tendency toward spirituality is, however, he also acknowledges that what we do with our genes is up to us. Thus, he supports the idea held in virtually all religious traditions: that spiritual practice leads to increased spirituality. It is important to note, though, that although it may be more "catchy" to refer to this group of genes as a "God-gene," especially in marketing a book, it is in fact more linked to spirituality, which may or may not be related to a belief in God.

It is not necessarily a problem from a Christian perspective if such a gene, or group of genes, exists. The idea that God has created humans with a desire for God has been a long-time refrain of Christianity, although there has not been a lot of speculation about whether there is a genetic basis to it. Whether or not this tendency toward spirituality is genetic or not is not in itself a problem; in fact, it could make perfect sense that this is how God created us. In addition to allowing for the possibility that spirituality can be within the reach of any of us, it also can account for some of the differences in spiritual sensitivity and practice among individuals.

Some have challenged Hamer's assertions, though, from the perspective of Christian theology. Hamer's definition of spirituality is problematic. Many people would not agree with his definition, which in its essence creates a strong separation and dichotomy between religion and spirituality. Religion is sometimes portrayed in a negative light when compared to spirituality, as though there is not a significant relationship between the two of them. Although one does not necessarily have to be religious to be spiritual, the two often do go together in individuals. And if spiritually is so defined as to exclude any real connection with religion, is it really any help in explaining belief in "God"? In addition, if there are spiritual receptors in the brain that can be "turned on" by drugs (which Hamer maintains), then is it something that is necessarily related to God? One could also question why we are not all genetically similar with regard to our spirituality, if indeed it is rooted in and "created" by God. The idea that God somehow predetermines one's inclination toward spirituality suggests that God may choose some in a way that God does not choose others. In other words, it brings up the Christian notion of predestination. Also, if there are genes for this tendency and we can influence them, should we? Depending on our religious persuasion, some might want to fix or remove this genetic tendency through genetic therapies, and some might want to induce it or increase it. Finally, we can also ask what the relationship of these genes might be to morality. For example, can individuals be blamed for not acting more morally by engaging in self-forgetful behavior if they are truly less prone to self-transcendence, and thereby self-forgetfulness, than others? Regardless of what one thinks about the concept of a "God-gene" or spirituality genes, Hamer's book is an example of how behavioral geneticists search for genetic predispositions for behavior. The fact that spirituality is a focus, however, should not be a cause for alarm, and studies in the future may clarify to what extent humans truly are genetically programmed to be spiritual, if at all.

CONCLUSION

What then, can we conclude about the possibility of finding a genetic basis for behavior and the implications of this for the human species? We would like to make several points. First, if we could indeed establish genes for behavior, there would be serious implications to and challenges for the Christian concept of God, human nature, and especially free will. If we are not truly masters of our destiny—if we are not truly capable of making free choices—then this affects our understanding of criminal behavior, moral responsibility, and even sin. It would challenge the very notion of what it means to be a human being, distinct from other species but also as made in the image of God. Second, a genetic marker and thus a predisposition for something does not make it set in stone; we still may be able to override or choose against our genetic orientations. And there is still the issue of the environment that needs to be taken into consideration as well. Third, if there is a genetic basis to human action, though, even in some circumstances, then we must take this into account in assigning moral responsibility and in our specific judgment of others. Perhaps individuals may not be as fully responsible for their actions as we have previously believed. Fourth, as the search continues for genetic markers for human behavior, there is general consensus among experts in many fields that we are a long way from finding candidate genes (and the mechanisms for their expression) for numerous behaviors. That being said, it is likely to be a search that continues for a long time to come. In fact, it may yield some very interesting results! Fifth, even though it may sell newspapers to announce a gene "for" something, most experts also agree that humans are not simply products of their genes; nurture (the environment) plays a big role in determining expression of certain genes, and many behaviors (as many diseases) are indeed multifactorial. Even with regard to genetic markers for diseases, which are a lot easier to find than those for behavior, we know that the expression of even a disease such as heart disease will be affected by lifestyle choices made by individuals. Thus, we must avoid genetic reductionism—reducing everything to our genes, especially in cases where we know that genes cannot solely be responsible. So, in answer to the question, "Did my genes make me do it?", at this point the most we can say is that although there may be a genetic basis to some behaviors, our genes should certainly not be considered destiny, and also should not be used to entirely avoid responsibility for the choices we make. Perhaps Jeff should be found guilty of murdering Bubba! But it seems that we need a lot more information on how genes affect behavior before we are even close to making such a determination.

NOTES

1. OMIM, http://www.ncbi.nlm.nih.gov/sites/entrez?db=omim (accessed December 8, 2008).

2. For a helpful discussion on this difficulty, see three essays in Erik Parens, Audrey R. Chapman, and Nancy Press, eds., *Wrestling with Behavioral Genetics: Science, Ethics, and Public Conversation* (Baltimore: John Hopkins University Press, 2006)—Kenneth F. Schaffner, "Behavior: Its Nature and Nurture, Part I," 3–39; Kenneth F. Schaffner, "Behavior: Its Nature and Nurture, Part II," 40–73; and Jonathan Beckwith, "Whither Human Behavioral Genetics?" 74–99.

3. S. Jonathan Singer, *The Splendid Feast of Reason* (Berkeley: University of California Press, 2001), 83–84.

4. Merriam-Webster On-Line Dictionary, 2005, Merriam-Webster, Inc., http://www.merriam-webster.com/dictionary/nurture (accessed June 8, 2008).

5. Eric B. Keverne and James P. Curley, "Epigenetics, Brain Evolution, and Behavior," *Frontiers in Neuroendocrinology* 29, no. 3 (June 2008): 398–412.

6. Richard Dawkins, *A Devil's Chaplain: Reflections on Hope, Lies, Science, and Love* (New York: Mariner Books, 2004), 106.

7. Avrum Stroll, *Did My Genes Make Me Do It? And Other Philosophical Dilemmas* (Oxford: Oneworld, 2004), 124–71.

8. James L. Gould and Carol Grant Gould, *The Animal Mind* (New York: Scientific American Library, 1994), 28–31.

9. Ibid., 46–53.

10. Dean Hamer, *The God Gene: How Faith Is Hardwired into Our Genes* (New York: Anchor Books, 2004), 94–95.

11. Keverne and Curley, "Epigenetics."

12. Edward O. Wilson, *On Human Nature* (Cambridge, MA, and London: Harvard University Press, 2004), ix–x.

13. Ibid., 84–85.

14. Richard Dawkins, *The Selfish Gene* (London: Oxford University Press, 1976).

15. Keverne and Curley, "Epigenetics."

16. *Nature*, PBS, http://www.pbs.org/wnet/nature/episodes/flight-school/the-man-who-walked-with-geese/2656 (accessed November 18, 2008).

17. Steven Rose, *Lifelines: Biology beyond Determinism* (New York: Oxford University Press, 1998), ix–xi, 11.

18. OMIM, #103780 (the # sign precedes the OMIM number because this is a polygenic trait) http://www.ncbi.nlm.nih.gov/entrez/dispomim.cgi?id=103780 (accessed November 22, 2008).

19. OMIM, http://www.ncbi.nlm.nih.gov/sites/entrez?db=omim (accessed November 22, 2008).

20. OMIM, 306995, http://www.ncbi.nlm.nih.gov/entrez/dispomim.cgi?id=306995 (accessed November 22, 2008).

21. Ricki Lewis, *Human Genetics: Concepts and Applications*, 7th ed. (Boston: McGraw-Hill, 2007), 141–46.

22. Alexis C. Wood et al., "High Heritability for a Composite Index of Children's Activity Level Measures," *Behavior Genetics* 38, no. 3 (May 2008): 266–76.

23. Mimi Bluestone, "Science and Ethics: The Double Life of Nancy Wexler," *Ms.: The World of Women* 2, no. 3 (November/December 1991): 90–91.

24. Marcus R. Munafo and Elaine C. Johnstone, "Genes and Cigarette Smoking," *Addiction* (2008): 1–12.

25. M. H. IJzendoorn, M. J. Bakermans-Kranenburg, and J. Mesman, "Dopamine System Genes Associated with Parenting in the Context of Daily Hassles," *Genes, Brain, and Behavior* 7 (2008): 403–10.

26. Carolyn M. Yrigollen et al., "Genes Controlling Affiliative Behavior as Candidate Genes for Autism," *Biological Psychiatry* 63, no. 10 (May 2008): 911–16.

27. Matt Ridley, *Nature via Nurture: Genes, Experience, and What Makes Us Human* (New York: HarperCollins, 2003), 6.

28. Ibid., 33–34.

29. Ibid., 126–27.

30. Ibid., 248.

31. Ibid., 267–69.

32. Leon Jaroff, J. Madeleine Nash, and Dick Thompson, "The Gene Hunt." *Time*, March 20, 1989.

33. Some recent important works are E. O. Wilson, *On Human Nature*; Matt Ridley, *Nature via Nurture*; and Ian G. Barbour, *Nature, Human Nature, and God* (Minneapolis: Fortress Press, 2002).

34. Heather A. Loy, "Is Our Fate in Our Genes? Behavior Genetics," in *Science and the Soul: Christian Faith and Psychological Development*, ed. Scott W. Vanderstoep (Lanham, MD, and Boulder, CO: University Press of America, 2003), 151–73. This is an excellent volume in general on the relationship between Christian faith and psychology with regard to human nature and the genetic basis for particular behaviors.

35. Robert Wachbroit, "Normality and the Significance of Difference," in *Wrestling with Behavioral Genetics: Science, Ethics, and Public Conversation*, ed. Erik Parens, Audrey R. Chapman, and Nancy Press (Baltimore: Johns Hopkins University Press, 2006), 235–53.

36. A very good collection of essays on the Roman Catholic approach to human nature can be found in Daniel N. Robinson, Gladys M. Sweeney, and Richard Gill, eds., *Human Nature in Its Wholeness: A Roman Catholic Perspective* (Washington, DC: Catholic University Press of America, 2006).

37. Leslie Stevenson, *Seven Theories of Human Nature*, 2nd ed. (New York and Oxford: Oxford University Press, 1987).

38. Stroll, *Did My Genes Make Me Do It?* 124–69.

39. For a very good volume on the effect of genetics on criminality, see Jeffrey R. Bottkin, William M. McMahon, and Leslie Pickering Francis, *Genetics and Criminality: The Potential Misuse of Scientific Information in Court* (Washington, DC: American Psychological Association, 1999).

40. Hamer, *The God Gene*.

41. Ibid., 213.

42. Ibid., 17–38.

43. Ibid., 63–78.

44. Ibid.

8

Don't Look at My DNA: Genetic Privacy and the Possibility of Discrimination

Annie's heart raced as she double-checked the list of questions on the job application. She was leaving active military duty and applying for a job with a major airline. Although she has been an Air Force pilot for many years, this time she was in a panic because she heard a rumor that for safety reasons they were contemplating genetic screening. Annie's family has a history of obsessive-complusive disorder (OCD), but she was not concerned until she learned that OCD may have a strong genetic component. Annie's OCD is very mild—she often double-checks cockpit instruments while flying. She is aware that some OCD cases develop into anxiety disorders and for that reason she is very careful to watch her stress level. On every other job application, including her application for military service, Annie did not mention her family history of OCD, but now she is in a quandary as to whether or not she should mention the family history. Annie views this as a no-win situation because she does know for sure how the GINA signed into law by President Bush on May 21, 2008, applies to her situation. She is concerned that even though her Air Force record is impeccable, her genetic information might somehow get into the hands of civilian airlines or the Federal Aviation Administration or both. If in good faith she acknowledges her family's history of OCD, yet emphasizing her impeccable military record, would the civilian airline still be justified in asking for a genetic test to rule out any predisposition for OCD? She could argue that her OCD is very mild, probably an acquired behavior, and for all practical purposes, double-checking flight instruments is a good thing to do while flying. The airline might rebut that although double-checking is appropriate in some cases, if double-checking becomes an obsession, it might take away from essential procedures with higher priority. The airline may also be concerned that the severity of her

symptoms may increase in stressful situations, particularly if she carries
the genetic marker. Should Annie indicate on her application that she has
OCD?

INTRODUCTION

Emerging genetic technologies have sparked a great deal of concern
regarding discrimination based on the genes one carries. Does Annie
risk discrimination for her family history of OCD? Annie is in a pre-
dicament that many of us are likely to face in our futures. How can
we best protect our DNA privacy, who will or should have a right to
it, and what protections can be put to place to protect citizens?
Annie's case is a bit more complicated because of her military situation. One of the growing concerns about genetic technologies in
general, however, is how to protect individual privacy. This is especially important when it comes to one's DNA. Significant questions
with serious implications have been raised. What is genetic profiling?
Should all citizens be required to submit a DNA sample and the
genetic information kept in a database? If so, who would have access
to such a database? What are some of the new technologies that are
likely to fuel this debate? How can we avoid genetic discrimination
with regard to healthcare and employment if one's DNA becomes
available to doctors, insurance companies, and employers?

During the early 1970s, African-Americans were often screened
for the sickle cell anemia alleles when applying for school or a marriage license. This practice led to employment discrimination and
caused Congress to pass the Sickle Cell Anemia Control Act of
1972.[1] At that time, genotypes were identified one gene at a time,
but today we can use automated laboratory procedures, computers,
and statistical methods to look at entire genomes. We can then
store this information in databases that can be accessed for forensics
and research, all with a corresponding potential for discrimination.
These concerns reached Congress again in 1995, around the time
scientists started to map the human genome, and relatively few
genetic tests were available. Nearly 13 years of debate ensued while
the human genome was mapped, with nearly 1,200 additional
genetic tests developed, and prospects growing for more personalized
therapies. The GINA was finally signed into law on May 21, 2008.
This law has been hailed as the "first major civil rights bill of the
new century" by Senator Edward Kennedy (D-MA).[2] GINA is

designed to protect individuals from discrimination by employers and health insurance companies if they carry genes predictive of future diseases, such as BRCA1/BRCA2 for breast cancer and HNPCC for colon cancer. It will take at least 12 months for heath insurance regulations to be developed, and another 6 months for employment regulations. GINA does not address potential discrimination for life insurance, disability insurance, and long-term care insurance, nor does it apply to members of the military.[3] The military is, however, taking steps to develop policies to protect members of the armed forces from genetic discrimination when determining benefits and upon their leaving the military.[4] We consider Annie's case more fully later in this chapter, but first we need to explore how it is even possible to "look at our DNA" and for what purposes. Then we consider the science behind forensic and medical profiling, the challenges we face in our efforts to protect genetic information, and we conclude with questions and recommendations.

GENETIC PROFILING: INFERRING FROM A "LOOK AT OUR DNA"

The technology exists to look at our DNA. We can even display our very own DNA profile as a work of art. *DNA 11* is a creative company that will send you a DNA sample kit as a first step in the creation of your very own DNA portrait for display as a wall hanging. For fun, and an additional fee, the company will identify several markers corresponding to specific genes that they claim are related to four areas: brain, love, intelligence, and hair color.[5] Your DNA profile as a wall hanging, with markers identified for specific traits, would surely be a conversation piece when hosting a party! But on a more serious note, DNA contains important information that scientists can use to infer a great deal about each of us, and the groups to which we belong (such as our ethnic group). Sir Alec J. Jeffreys is professor of genetics at Leicester University in the United Kingdom; he developed a technique in the early 1980s to identify individuals by examining variations in their DNA. His discovery was first put to use to confirm British citizenship for a boy whose family originated in Ghana, but the tremendous forensic value of this technique was quickly recognized when it was made famous as a tool to identify a man who killed and raped two young women in Leicester. Thus the technique was first called "DNA fingerprinting," and although this term persists today, it does not reflect the wide range of technologies and applications that have evolved for forensic, medical, genealogical, and other purposes. The term genetic profiling is used here to represent a process that includes collecting, examining, storing, and analyzing the DNA to infer information about an individual and or a group (typically a population). To understand the value of DNA for profiling, whether it is for forensics or some other purpose, it is first necessary to illustrate the fact that genetic information is arranged in both coding and noncoding sequences, and that these sequences contain observable variations in DNA structure. This helps us understand how scientists are able to observe our unique variations by cutting our DNA and then sorting the fragments.

Genetic information is coded in a chemical language that all forms of life can read and express, and DNA carries that information coded in sequences of the four chemical bases A, C, G, and T. These letters, like our alphabet, can be assembled into meaningful sentences called coding sequences; this DNA functions as genes for the expression of traits. Genes can, however, vary slightly in form and still code for a functional protein, somewhat like a sentence with a few misplaced words that do not alter the overall meaning. For example, one of the common genes associated with blood type has three alternate forms, alleles A, B, and O. In contrast, nearly 3 million letters or base pairs in the human genome (about 0.1 percent) appear to form meaningless sentences, noncoding sequences of DNA, but nature's editor (natural selection) does not make the effort to cut them out. Both coding and noncoding variations in DNA sequence are called alleles: there are variations in coding DNA sequences for observable traits such as the blood-type alleles, and there are noncoding DNA sequences within genes and between genes. The noncoding alleles are the so-called junk DNA that may have functions which are not yet understood, but when it comes to forensics and genetic epidemiology, this DNA is definitely not junk. Absent the forces of natural selection, variations that occur in noncoding DNA sequences are not selected out, and therefore accumulate and persist in families and populations over time. For this reason, looking at an individual's DNA for variations is most valuable when compared to variations in populations, but before we can do that, we need to understand how it is possible to observe variations in a sample of DNA.

Scientists are able to identify variations in DNA, and commonly characterize these variations as either repeating chemical-base sequences or changes in base sequence.[6] Repeating chemical-base sequences are like duplicating letters or words in a sentence; they are typically differentiated as short tandem repeats (STRs; typically 2–6 bases repeating 7–15 times), and variable number tandem repeats (VNTRs; typically 7–25 bases repeating 50 times). The shorter STRs can be more useful than the longer VNTRs when DNA samples are too small or degraded, because it can be difficult in some cases to collect DNA of sufficient length for amplification and analysis.[7] Alec Jeffreys used VNTRs to develop DNA fingerprinting, but today DNA profiling includes a wide range of new technologies and relies heavily on STRs. It is difficult to work with small samples of DNA, but biologists can amplify a sample of DNA using an enzyme (DNA polymerase) that occurs naturally in cells for the replication of DNA before cell division—a photocopy-like process for DNA. This commonly used technique developed by Kary Mullis is called polymerase chain reaction (PCR). DNA profiling typically begins with a sample of DNA that is amplified as necessary and treated with one of the many types of restriction enzymes found to occur naturally in bacteria. How this process works illustrates how nature has provided us with the tools to seek out and identify specific variations in DNA. Bacteria use restriction enzymes to fend off viruses that depend on their molecular machinery to reproduce. The process begins when a virus inserts its DNA into the bacterium for construction of more viruses via bacterial systems, but the bacterium fights back with restriction enzymes that cut the intruder's DNA into pieces. It just so happens that bacteria contain restriction enzymes that cut DNA at very specific base sequences, and these enzymes are specific to each species. Imagine that the previous sentence was DNA, and that we used a hypothetical restriction enzyme—the cut and paste function on our computer—to recognize the letters "act" (located in the word bacteria) as a place to cut our DNA sentence. In nature, the cutting is not random; each type of restriction enzyme recognizes a specific sequence on the DNA for cutting. Thus, for our example we are restricted to cutting only the region of the sentence with the letters "act," and if we did this over and over again, the result would always be two fragments, a short fragment ("It just so happens that bact"), and a remaining longer fragment ("eria contain restriction enzymes that cut DNA at very specific base sequences, and these enzymes are specific to each species").[8] But how are these fragments sorted for observation?

Jeffreys used restriction enzymes to cut DNA into fragments, and then sorted those fragments by size using a technique called electrophoresis. This method uses a special type of gel in a tray with the DNA fragments placed at the end of the tray opposite a positive electrode. Since the chemical nature of DNA causes it to have a very small negative charge, the fragments migrate through the gel toward the positive electrode, with smaller fragments moving more easily and therefore a greater distance than the longer fragments. The different size fragments are called restriction fragment length polymorphisms (RFLPs). The DNA fragments could then be stained and viewed as bands in the gel, but this is not adequate for forensic work today. Instead, the DNA is prepared in such a way as to allow it to combine with labeled DNA probes (we describe the process later); this allows for viewing the DNA alleles as bands that typically look like a barcode. Consider the following analogy to illustrate how this process can cut and sort DNA to identify individuals. Imagine that each of us wrote a short story about our childhood, and that the words in the story represent base sequences in a sample of our DNA. Next, using a scissors (like a restriction enzyme), you cut the sentences in your story wherever the word "is" occurs. The result would be fragmented sentences that can be sorted by size. If we were to then create a bar graph showing the number and sizes of the fragments for each of us, it is possible that each of us would have a unique fragmented sentence profile. But if not, we could do a second cut wherever the word "and" occurs, and then each of us might have more unique fragment profiles. This process could be repeated until we have a unique fragmented sentence profile for each participant. The process described in this analogy is similar to the procedure that is used by biologists to identify the short (STR) and long (VNTR) fragments that are a result of STR and VNTR variations when DNA is cut with restriction enzymes. However, these are not the only types of variation observed in DNA.

A second type of variation in DNA that biologists can identify is base-sequence changes. About 95 percent of sequence variation is believed to take the form of what biologists call SNPs.[9] These variations can occur within the coding or the noncoding DNA sequences. Variations in coding sequences (single and multiple) are responsible for alleles, or alternate forms of genes for various traits. For example, the most common sickle-cell anemia allele contains a single chemical-base variation that results in glutamic acid being replaced by valine in the hemoglobin protein.[10] SNPs in noncoding DNA sequences are believed to be of no consequence, but SNPs in both coding and noncoding regions are informative. SNPs are like typographical mistakes; they can reveal a tendency to transpose letters on the keyboard, and they can indicate a natural error rate in our typing. For example, the SNP associated with the sickle-cell trait is more abundant in populations exposed to malaria; this is because the phenotype interferes with the malaria-causing parasite's life cycle (an example of what biologists call a balanced polymorphism). In the sickle-cell trait, natural selection has stabilized the abundance of an SNP in the coding region of a gene. SNPs in noncoding regions may also occur as spontaneous mutations in individuals, and then pass on to descendants where they persist over time without natural selection editing them out of the genome for the population. Thus, SNPs can be used to identify genotypes associated with phenotypic traits, and they can be used to identify individuals and describe their genetic heritage. SNPs and other genetic variations tend to travel together, generation after generation, depending on how closely they are linked on a chromosome (recall that recombination in meiosis shuffles linked genes). Variations that are closely linked travel together and are called haplotypes (short for haploid genotype) by biologists. The International SNP Consortium is an organization to assist scientists and funding agencies in cataloging and identifying these genetic differences in humans. This public database is part of the developing HapMap for humans; it includes common variations where they occur in the human genome and describes frequencies for their distribution in and among populations of the world. This type of data is then available for scientific research.[11] This pooled information is a valuable source for the development of pharmaceuticals and medical treatments, and serves as an example of what is commonly called a biobank, which we discuss later in this chapter.

To appreciate the value of biobanking genetic information, we first illustrate how genetic variations can be used in genetic profiling. The technical capacity to use variations in DNA to identify one person among many has revolutionized forensics. Before DNA analysis was possible, criminologists used other substances of biological origin that can vary in form, such as proteins that are coded for by genes, but biological substances other than DNA exhibit much less variability. Human blood type is one example, and obviously, blood is very common at most crime scenes. The three common alleles (A, B, and O) for blood type in humans (recall that chromosomes come in pairs and a person can carry only two alleles at a time) can combine to yield six genotypes (AA, AB, BB, AO, BO, and OO) and four phenotypes because of the existence of dominant and recessive alleles (recall from chapter 3 that the A and B alleles are codominant). Today, crime labs generally use 13 or more loci when they compare their samples to state and national databases. Loci that are important in forensics are not dominant or recessive, but they do vary according to the size and sequence of the DNA. The greater the variability at a given locus, the greater the discernment when several loci are considered in the analysis of DNA from a crime scene.[12] But before DNA can be analyzed, it must be gathered—carefully!

Nucleated cells in tissues and fluids contain the DNA that is necessary for analysis. Many of us enjoy crime scene investigation (*CSI*) television shows where we observe DNA samples carefully gathered in ways that are usually much more exciting than using a cotton swab to gather cells from inside the cheek. Collecting biological evidence such as blood, semen, saliva, hair, teeth, bone, and other tissues and fluids must be done without contamination at a crime scene. Biological evidence is typically collected from a surface or material by swabbing, scraping, or cutting. To be useful, the DNA must then be isolated using laboratory procedures to yield strands that are large enough for analysis (typically more than 20,000 bases). Typical laboratory procedures include organic extraction of DNA from white blood cells in liquid whole blood, and inorganic methods to extract DNA from a mixture of proteins. In the field, the *FTA* card can be used for DNA preservation until future laboratory analysis. Blood combined with other materials can also be collected. Promega Corporation has developed a magnetic resin procedure that is very effective at gathering substances containing DNA from materials such as denim, leather, and soil; lifted stains are heated in a chemical buffer to rupture cells for subsequent removal of debris. The resulting solution is then mixed with a magnetic resin containing particles that attract genomic DNA for collection using a magnet. In all cases, the DNA must be "quantified" in the laboratory to verify that it is not degraded, and of sufficient size for analysis.[13] Collecting and analyzing DNA must be done by experts to increase the chances that it will be admissible as evidence in court, but it is possible for individuals to purchase kits containing *FTA* cards to collect DNA. These cards are marketed to laboratories, coroners, funeral home directors, private investigators, those interested in child identification, and schools, to name a few.[14] For about the price of a meal, one can collect a sample from a few friends and then store their DNA as a keepsake, but it might be cheaper to pull a few of their hairs. All of this should be done with their consent, of course! You would have to make sure that each hair includes a follicle, the cell where the DNA is housed, and then take steps to ensure that it does not degrade completely, or get lost. But how small can the sample be?

Technologies to collect and amplify DNA for analysis have improved dramatically in the last 10 years. DNA for analysis is typically isolated from chromosomes or mitochondria. The process begins with biochemical techniques to rupture the cells and separate the DNA from cellular materials for subsequent identification. Early techniques looked at VNTRs, but it is not always possible to extract complete sequences of DNA. If the DNA is fragmented, then STRs are amplified for analysis, and if the DNA is extremely damaged, then mitochondrial DNA is used. It remains to be seen how small a DNA sample can be before it becomes unreliable; in all cases, reliable identification depends on the molecular analysis of a DNA sample. This enables forensic experts to compare samples of interest, such as those found in the JonBenét Ramsey case, where people's lives are affected if the evidence is not available. A technique called touch DNA analysis recently cleared John and Patsy Ramsey of public suspicion that they may have been involved in the killing of their daughter, JonBenét, more than 10 years ago. DNA recently collected from the waist area of her long-johns was analyzed and found not to match the DNA of her parents. This touch DNA was then observed to also be mingled in a sample of JonBenét's blood found elsewhere on her underwear. Thus, 10 years later it is possible to clear the Ramseys of suspicion with this new technology.[15] The technical term for this procedure is low copy number DNA profiling, and it is important to note here that concerns have been raised regarding the reliability of using trace amounts of DNA.[16] If DNA evidence is carefully gathered in the field, and examined using standard procedures in the laboratory, then DNA profiles can be used to include or exclude suspects using allele frequencies within populations.[17]

Crime labs generally use 13 or more loci when comparing crime scene samples to state and federal databases. Each locus is on a different chromosome and has many possible alleles; dominance does not apply because these are noncoding sequences. The greater the variability at a given locus, the greater the discernment when several loci are considered together in the analysis. To illustrate here, we use only 2 of the 13 loci in the Combined DNA Index System (CODIS): a noncoding region of the human thyroid peroxidase gene (the TPOX locus) and a noncoding region of the human α-fibrinogen gene (the FGA locus). The TPOX locus is a repeating GAAT base sequence on chromosome 2 with 8 common allelic forms to yield 36 possible genotypes (the math is 8+7+6+5+4+3+2+1=36 possible paired combinations). The FGA locus is a repeating CTTT base sequence on chromosome 4 with 14 allelic forms to yield 105 possible genotypes. Sex is determined by examination of the DNA sample for variations within the noncoding sequences of the amelogenin gene located on the X and Y chromosomes (this gene is responsible of tooth bud development).[18] Public crime labs with the appropriate CODIS software can retrieve and create authorized DNA profiles in the National DNA Index System (NDIS) maintained by the FBI. The CODIS system relies on two indexes: a convicted offender index, and a forensic index that contains biological evidence retrieved from crime scenes.[19] DNA profiles can be used to exclude or include a person when considering biological evidence at a crime scene, and a DNA sample from a crime scene can be compared to profiles in the convicted offender index. Recall that the TPOX locus has 36 possible genotypes and that the FGA locus has 105 possible genotypes. How do we estimate the chances that an individual will carry any given genotype? We use the product-rule calculation to predict the chances of two independent events occurring simultaneously. This calculation multiplies the chances of each event occurring independently to obtain a probability of the events occurring together. For example, the probability of observing three tails in a row when flipping a coin is one-half times one-half times one-half or one in eight chances (a frequency of 0.125 or 12.5 percent). The product rule can also be used to predict the chances that an individual would carry a particular combination of alleles used in the CODIS system. For example, the chances that an individual carries a given genotype for both the TPOX and FGA loci are 1/36 times 1/105 or 1 in 3,780 chances (a frequency of 0.00026 or about 0.03 percent). This is probably not enough for a conviction, but the chances become smaller and smaller with each additional allele factored into the multiplication. Commercial equipment and kits are available to assist laboratories in their analysis of loci, and if 13 loci are considered, then the chances are about one in several trillion that two people would have the same genotype. These conventional methods for examining variations in DNA and then banking the information in databases for use continues, but the development of microarrays has also facilitated the rapid generation of databases containing genetic information by other sciences, such as epidemiology and genealogy.

BANKING GENETIC INFORMATION

Genetic epidemiology is a science that considers the interplay of hereditary and environmental factors that affect health in families and populations. Genealogical studies use genetic information to examine human origins and to assist families in constructing their family trees. These sciences, including forensics and many other disciplines, have benefited from microarray techniques and computer information technologies that enable scientists to gather, store, and analyze large amounts of genetic information in families and populations. This is because microarray technologies make rapid genotyping of individuals possible.

In this section, we begin by briefly describing how microarrays work, and then we examine the biobanking of genetic information. The first gene chip (a microarray) was developed by Affymetrix in the early 1990s. As the name implies, these chips have features in common with manufactured microprocessors, but these gene chips are made of glass instead of silicone. They are small coin-sized wafers that contain millions of DNA probes which are added one nucleotide at a time.[20] Today, Affymetrix includes in its large product line a MyGeneChip. This is a custom GeneChip containing probes for gene expression and genotyping of human, animal, plant, and microbial genomes.[21] A primary component of the microarray is a DNA or RNA probe; we do not detail the differences here because it sufficient for our purposes to simply refer to these molecules as nucleotide sequences. But some understanding of basic biology is necessary to understand how probes work.

To begin, recall that it is possible to make millions of copies of nucleotide sequences (cloned DNA and RNA). To do this, a new nucleotide sequence is built up from another nucleotide sequence, one base at a time, according to strict pairing rules for chemical base pairs (for example, G always pairs with C). Scientists rely on these strict pairing rules when they synthesize a sequence of nucleotides for use as a probe to seek out and combine with a complimentary sequence in a sample—this is a powerful "search engine." It is similar to using the find function on your word processor to locate a word or phrase in your document. Probing requires that we have a single-stranded nucleic acid molecule so that the chemical bases can line up; the more chemical bases in common, the more fidelity in the pairing process (called hybridization). For example, if I opened the find function on my word processor, and typed in the phrase "Annie's OCD is very mild" and pressed enter, the computer would act like a probe and locate that phrase in this chapter. Coding and noncoding nucleotide sequences of interest can be identified using probes that are marked with a beacon of some type (fluorescent or radioactive); this allows scientists to identify nucleotide sequences for alleles of interest in a sample that are a complementary match to the probe. Thus, it is possible to manufacture microarrays with thousands of probes labeled for specific chemical base sequences robotically applied to chips at marked locations (almost like a dot matrix printer), and now automated systems exist to process genetic samples. Affymetrix recently introduced the GeneTitan, and the company promotes this combined microarray processor and scanner as a way to allow scientists to spend much less time acquiring data.[22] This technology has made genotyping easy, inexpensive, and fast for scientists and consumers. For example, for less than 100 dollars you can purchase a genetic paternity test kit with much more specificity than the old blood-type method.[23] Thus, it should be no surprise that scientists and nonscientists are awash with genetic information for humans, and many other forms of life. Where does all this data go, and how is it used? It is banked!

Scientists have been gathering and organizing biological data for a long time, and today, genetic information is more abundant than ever. Genetic information can be stored on paper and in digital format; it can also be maintained in tissue cultures or host organisms (living libraries). For example, DNA from any organism can be maintained on YAC in what is called a genomic library (chromosomes carry genes on a protein scaffold and replicate with each cell division). It takes more than 3,000 YACs to store the entire human genome.[24] The human genome was originally sequenced using a collection of DNA strands from several individuals, and although this has provided a complete sequence of possible genes and their order, it does not account for all of the variability within coding and noncoding sequences. In fact, today it is easier to gather the information than it is to understand what all the information means! Many scientists store the data they gather on public databases so that other scientists can help in the analysis. One example is the NCBI Web site that contains genome sequence data in GenBank, and medical research information on PubMed.[25] Scientists are browsing the entire genomes of plants, animals, and microorganisms in an effort to understand the form and function of genetic information, evolutionary history relationships, and how genes and the environment interact. It is still very time-consuming and expensive to sequence a person's complete genome; this is a phenomenal task because there are about 3 billion bases to sequence. To give you an idea as to how big a number this really is, it has been estimated that it would take 9.5 years to read each base letter in a person's genome out loud without stopping.[26] Until recently, the cost for a complete sequencing was hundreds of thousands of dollars, but in the second quarter of 2009, a start-up company called Complete Genomics plans to offer a complete sequencing for as little as 5,000 dollars.[27]

Forensic, genealogical, and epidemiological studies rely on databases containing DNA sequence variations and one thing is for sure: it is likely to be less expensive and easier to rapidly sequence one's entire genome in the near future. Recently, a team of scientists reported sequencing the entire genome of an individual, James D. Watson, with a relatively high degree of accuracy in a two-month time period, at one-hundredth the cost of conventional methods and with fewer lost sequences. (Recall that James D. Watson, along with Francis Crick, discovered the structure of DNA.) Watson's genome was sequenced using an automated sequencer. It was then compared to the human reference genome at NCBI by sequence alignment; a tool for this purpose is the Basic Local Alignment Search Tool (BLAST). Qualified matching sequence alignments were then compared to sequenced fragments to distinguish sequencing error and genomic variation in the sequenced genome. Nonmatching sequences and low-quality partial matches were then pooled for sequencing using a standard genome assembly procedure. Watson's DNA sequence was then labeled and placed on a Gene-Chip manufactured by Affymetrix for independent laboratory analysis. The research team for this study also reported their efforts to protect the privacy of Watson's genetic information with a full description of steps taken as part of the paper.[28] We discuss similar steps to ensure privacy later in this chapter, but first we continue discussing banking of this information.

Biobanks can be broadly defined as repositories for biological specimens and information, but for our purposes here, biobanks (or more specifically genebanks) are population-based genetic databases. Two common genetic databases have already been mentioned: NCBI, maintained by the NIH for access using various methods; and NDIS, maintained by the FBI for authorized access using CODIS software. Public health initiatives to facilitate epidemiological studies have resulted in the development of several national biobanks. For example, one of the first to be established is the Iceland Health Sector Database (IHSD). This is a government-sponsored genebank managed since 1998 by a private company, deCode, and it includes more that 70,000 genetic samples. Consent to participate by individuals is presumed; this is a so-called "opt-out" system, and third-party encryption is used to protect data. A second example is the United Kingdom genebank, which includes DNA, medical records, and lifestyle information. This biobank is sponsored by UK's Medical Research Council, the Wellcome Trust, and the Department of Health, and it includes data from more than 500,000 volunteers. A third example is the Estonia genebank, which includes genotypes, medical records, and genealogical data for more than a million consenting participants. This biobank was established by Parliament and maintained by the Estonian Genome Project Foundation and its commercial arm, EGreen Inc., a private company that intends to compete in the global pharmaceutical market.[29] Biobanks vary widely in size and sponsorship, but a common theme is emerging—privacy requires vigilance. The United Nations Educational, Scientific and Cultural Organization (UNESCO) adopted an International Declaration on Human Genetic Data to address the proliferation of biobanks and concerns about use and sharing of this information, while acknowledging the potential value of information-sharing.[30] Although these banks vary widely in sponsorship, they do appear to have a common goal: to assist epidemiologists in their search for genes that are associated with complex diseases to improve the human condition.[31] But several questions remain. Is participant consent adequate? Who owns the data? How might the data be shared and used as technologies change? Biobanks could provide the necessary information for genome-wide analysis as scientists seek to improve the human condition, but can we protect the individual from discrimination as a result of this genetic information?

Protecting privacy, while making the data available for forensics, genealogical, and epidemiological studies, is difficult. Forensic databanks include genetic data that is normally gathered at crime scenes and admitted into the legal system, whereas access to the data is limited to authorized searches within databanks for the purpose of excluding or including individuals as suspects (for example, CODIS). Genealogical banks use similar methods, usually with a person's consent, to exclude or include individuals in families, groups, and populations, and even to trace migrations. For example, all individuals are invited to voluntarily participate in the Geno-graphic Project sponsored by the National Geographic Society, IBM, geneticist Spencer Wells, and the Waitt Family Foundation. The project is designed to use DNA samples from volunteers all over the world in an effort to chart human migration patterns out of Africa. Each DNA sample collected is examined for genetic markers on the Y chromosome in males, and mitochondrial DNA in females, and then used to chart human migration patterns.[32] Mitochondria are passed from mother to both male and female children, and scientific evidence suggests that all living humans today carry mitochondria that are descendent from a so-called "mitochondrial Eve" living in Africa about 150,000 years ago; there must have also been a "Y chromosome Adam."[33] It is possible to figure this out because DNA markers that characterize a group's DNA are like a language containing acquired local and regional dialects, but one advantage of DNA is that natural mutation rates in DNA can be used to estimate when markers arise. This property of DNA has enabled scientists to chart time along with migration patterns. These authors participated, and we were surprised that our ancestors shared common migration patterns. One of us consented to release his data, and several months later he received an e-mail from Family-TreeDNA.com stating that "an exact 12 marker match has been found between you and another person in the Family Tree DNA database."[34] He was surprised to learn of yet another living relative. Thus, with individual cooperation and the sharing of information, we might find long-lost relatives and gain a more comprehensive view of human origins. But what are the chances for personal identi-fication if someone has access to this type of information? It is com-mon practice to use names when developing a family tree, but when constructing family trees with genetic information on searchable databases, it is more appropriate to use a less direct form of identifi-cation (i.e., pin numbers). If the original DNA sample is in storage, then future access may provide more additional data as sequencing the person's entire genome becomes feasible. Thus, concerns are likely to arise as new technologies facilitate further use of existing DNA samples and data. This will require the development of ways to ensure that previous contributors of DNA samples and data are given the opportunity to consent again for an additional use of their DNA data, or have the option to withdraw their data. Safeguarding privacy becomes a complex challenge. Once a person's data is out there, it is difficult to tell where it is, who will use it, and for what purpose.

Genetic databases have greatly enhanced efforts to solve crimes, to identify disaster victims, to trace human origins, to connect fami-lies, and to research ways to match therapies and genetic predisposi-tions. Access to databases containing genetic information is essential if these efforts are to be successful. As new technologies arise, agencies are acting to protect privacy. Genome-wide associa-tion studies rely on databases, but recent concerns about the privacy of individual data have caused Wellcome Trust (the UK biobank mentioned earlier), the NIH, and the Broad Institute to remove "potentially sensitive" genetic data from their Web sites and to require scientists to apply for access.[35]

This action was prompted by a significant piece of work published in the on-line open access journal, *PLoS Genetics*, on August 8, 2008. Nils Homer and a team of scientists demonstrated that it is possible to identify the presence of an individual's DNA in a com-plex mixture of DNA from other persons, even when the individu-al's DNA is less than 1 percent of the total. This groundbreaking work was achieved by examining hundreds of thousands of SNPs using microarray technologies and statistical analysis. Referring to genome-wide association studies, the authors state the following: "Although counter-intuitive, our findings show a clear path for iden-tifying whether specific individuals are within a study based on sum-mary-level statistics."[36] Clearly, this is an example of how, as these technologies develop, it is possible that data that were once thought to be secure are no longer secure. Two recent profiling tools under development illustrate this point and may someday surprise us with amazing new possibilities. First, efforts are under way to develop software programs that can use DNA to generate facial profiles.[37] This brings to mind the task of a sketch artist working for a police department, but in this case a DNA sample is the witness. The second technology is LC DNA profiling, which was applied in the JonBenét case. If these technologies were to converge as DNA "traces to faces" profiling, then one can imagine that DNA samples could be used to generate a host of phenotypic traits to help identify individuals. One danger is that a large group of people with similar characteristics might become suspect.[38] We should want to embrace technologies that have great potential to make our world a safer place, to reunite families, to personalize medicine, to explain human origins, and possibly to even enhance our understanding of life on planet Earth. But can we do this while simultaneously avoiding discrimination against individuals or groups, and protecting individual privacy?

GENETIC INFORMATION: IS OUR GLASS HALF EMPTY, OR IS IT HALF FULL?

Perceptions that we hold regarding genetic information make it different from other types of information. On the one hand, we perceive our genetic information as private, possibly because it has the capacity to reveal information about our past, present, and future. On the other hand, we perceive genetic information as having the capacity to reveal to us our limitations. We return to the introductory case to illustrate this point. Annie was aware of her family's history of OCD, but she was not concerned until she learned that OCD may have a strong genetic component. For Annie, the glass suddenly became half empty; she was afraid that her future as a civilian pilot would be limited. Why is perceiving a family history so different from knowing one carries a gene associated with a trait? In Annie's mind, a future limited by a family history may not be as deterministic as having a gene "associated" with OCD. Taking this line of reasoning one step further, it is even more limiting for those who test positive for a gene if it is said to "cause" a particular disorder. We have a great deal to learn about how genes associated with traits, particularly complex traits such as OCD, can interact with the environment to produce the phenotype. A further complication is the so-called therapeutic gap. On the one hand, we may be much more likely to see the glass as half empty when a cure is not known for a genetic predisposition or trait, such as Huntington disease. On the other hand, we may see the glass as half full if preventative measures exist, such as in the case of colon cancer.

The focus here has been on the individual, but what about the needs of the group? For example, we may desire protection for Annie from job discrimination as a pilot, yet we also desire protection for those who travel by air. Thus, there might be compelling reasons to solicit and act on an individual's genetic information for the needs of the group. This information is of exceptional value; when it is compiled and analyzed as aggregated data, it can provide benefits to the group. This information can assist the group's legal system, facilitate research to enhance therapies, explain our origins, and reunite families, to name a few. As individuals, some of us see the glass as half full and are willing to take a risk and release our data for the benefit of all, yet others of us see the glass as half empty and are not willing to take the risk. Thus it is safe to say that genetic information is exceptional, and therefore we must take steps to offer a greater degree of consent when asking individuals to release that information. What steps need to be taken?

There is a need for serious dialogue regarding the use of genetic information in whole-genome research (using genetic databases, and sequenced genomes). Recently, Timothy Caulfield of the Health Law Institute at the University of Alberta, Canada, led an interdisciplinary team of scientists and ethicists to develop a consensus statement on ethical considerations for whole-genome research. Their recommendations are extensive and warrant serious consideration by institutional review boards when dealing with genetic information. For our discussion here, we draw on issues raised in several of their recommendations regarding the use of an individual's genetic information: consent, withdrawal from research, results, and public data release.[39] Fundamental to the challenges when dealing with DNA information and informed consent is the fact that DNA, given our current state of understanding, appears to be different from most forms of information. It takes a great deal of scientific understanding to even comprehend how special DNA is, and thus, it is difficult to ensure that one is informed fully. Even those supposedly "in the know" cannot fully anticipate the outcomes, certainly when the technology is changing so rapidly. The concept of a social security number (SSN) may serve to illustrate several of the issues surrounding the collection and handling of DNA samples and genetic information.

We received the DNA that we have from our biological parents, and although it is possible that some of us may have had our DNA sampled or tissues stored, for most of us our DNA is our own private information. This is not the case for a SSN, because when we are assigned a social security number, the data to identify us are already in government hands. Nevertheless, we do strive to protect our SSN from those that might desire to use it for the wrong reasons. We do, however, share the number with employers, banks, licensing agencies, and other entities. The very act of sampling one's DNA is much more complicated than writing down your SSN. Accurate DNA sampling requires protocols and quality controls; thus there should be ample opportunity for the entity accepting your DNA to properly inform you as to how the DNA will be processed, analyzed, and stored, and for what purposes. Assuming one understands the science, this still may not be enough information to make an informed decision. As we have seen earlier, it is very likely that DNA will become an even more precise profiling tool in forensics and medicine, and if we add genetic screening and modification to the mix, it is nearly impossible to anticipate all future possibilities. Thus, one could argue that it is impossible to inform one fully of the potential uses for their DNA.

Imagine that you placed your SSN in an e-mail and sent it to a trusted friend. This is could be compared to taking a cotton swab in a DNA collection kit, gathering your cheek cells, and then mailing them to a trusted company, or some other entity, to have it analyzed. Can you be absolutely sure that your SSN and DNA will remain confidential in each of these cases? The answer is no! Your friend might accidentally forward the e-mail with your SSN to someone else, or someone might hack into her computer, or your friend might suspect you of avoiding taxes and turn you in; the possibilities are endless.

DNA carries additional risks; typically, the sample is examined for specific loci, and then it is either stored or discarded. If only the data are stored, what is going to prevent the company from sharing the data? If your DNA sample was also stored, and whole-genome sequencing technology becomes readily available, what is to stop the company from sequencing your stored sample and using the information in some other way? Clearly, informed consent needs to consider these possibilities, and those who approve informed consent protocols need to make every effort to see that research participants have the opportunity to consent again for additional uses of their DNA and the information, or even withdraw from the process. To offer this opportunity will require ongoing monitoring of the science and technologies involved, and a concerted effort by researchers and those in policy-making arenas to decide at what point opting out is no longer a feasible option. Assuming we do find ways to address these concerns, then the next issue is the sharing and use of the information.

We have an expectation of privacy when our information is shared by others, and we anticipate benefits when sharing our information. Most of us have shared our SSN, and when we inquire about our credit card balance by phone, it is not uncommon to hear a request for the last four digits of our SSN. If we initiated the call, then in most cases we are willing to release these digits, because we realize that it is necessary to link us to a private database including our SSN and all kinds of financial information. Thus, as it is with our SSN and financial information, it is necessary to find some way to link, in a safe and secure way, our genetic information and demographic information about us (e.g., criminal convictions if applicable, medical and genealogical information if provided). If databases are to be useful tools for researchers, then they must be large enough for analysis, representative of populations, and searchable using reliable and valid methods—all of these desired attributes are pressure for the sharing of data. But before releasing our DNA we should be informed as to what will happen to our original sample, because as the technology improves, so does the utility of our sample. We should know if our sample and or the data are to be shared with other databases, and if so, for what purpose. With each additional use we should be given the option to consent again or withdraw our information. For example, complete sequencing may become so easy and inexpensive that genetic information in samples on file may be extracted, but to what end? We might reconsider when given this information. Clearly, we are in the midst of more questions than answers in light of a technology that is growing faster than our ability to anticipate the consequences. Can we as a society make an informed choice?

CONCLUSION

Our DNA contains information about our past, present, and future. Clearly, this is information that we want to protect. We discussed how scientists look at our DNA to examine the minor variations that makes each of us unique. We examined how scientists can probe this information for traces of information that is proving to be very useful in forensics, medicine, and efforts to better understand our origins. Some of us may see the glass as half empty, others of us may see it as half full, but all of us must participate in the public debate if society is to make wise choices.

U.S. Supreme Court Justice Stephen Breyer offers some words of wisdom on the issue of genetic privacy: "To act coherently, we in the law must be able to have a sense of the likely social and economic impact of our choices 'before' we act—so that our decisions are grounded in realistic predications of what science will do, and not fanciful predictions of what science might do."[40] Breyer reasons that because the Supreme Court is the "court of last resort," it is essential that there be more public discourse on the issue of emerging genetic technologies, particularly when it comes to the legal system, genetic diagnosis, and the ownership of genetic information. Breyer predicts that because of DNA evidence, trials will become more reliable, and past convictions will be reconsidered. He acknowledges that genetic diagnosis will increase, but accuracy is important, and access to this information may conflict with privacy issues. We are not accustomed to knowing this kind of information in advance, so counseling may be necessary and issues of financial responsibility will arise. Breyer suggests that the policy debate on "environmental contributors" to disease may also change. We have tolerated many carcinogenic substances such as barbecued foods, gasoline, and pesticides because the small risks are hard to remove and we are not sure who will be affected. But what if we have an idea who will be affected? Thus he raises a justice issue. "What will happen if certain of those products create large risks for a few individuals whom we can identify in advance? How can we, how should we, selectively regulate their exposure?"[41] Breyer emphasizes that answers to these questions require both specialized and interdisciplinary knowledge, and depend on our anticipating uncertain social consequences. It should not be surprising that these policies are currently being developed in areas of law where revision and reversals are easier to make, but in the long run, and after much debate, it may be necessary to consider whether or not fundamental changes to the Constitution are necessary to address these issues.[42]

NOTES

1. Susannah Baruch and Kathy Hudson, "Civilian and Military Genetics: Nondiscrimination Policy in a Post-GINA World," *American Journal of Human Genetics* 83, no. 4 (October 10, 2008): 435–44.

2. Kathy L. Hudson, M. K. Holohan, and Francis S. Collins, "Keeping Pace with the Times: The Genetic Information Nondiscrimination Act of 2008," *New England Journal of Medicine* 358, no. 25 (June 19, 2008): 2661–63.

3. Ibid.

4. Baruch and Hudson, "Civilian and Military Genetics."

5. DNA 11 Art, http://www.dna11.com (accessed November 25, 2008).

6. To avoid confusion, it should be noted that the terms "base" and "base pair" are commonly interchanged. To be precise, base pair takes into account the fact that DNA is normally double-stranded, and if the sequence of one strand is GATCTA, then the other strand sequence is CTAGAT. Forensic applications do use single strands of DNA.

7. Lawrence Kobilinsky, Thomas F. Liotti, and Jamel Oeser-Sweat, *DNA: Forensic and Legal Applications* (Hoboken, NJ: John Wiley & Sons, 2005), 18.

8. This analogy can be carried one step further by mentioning that the freshly cut sites have sticky ends and will recombine if conditions are appropriate. This property enables scientists to cut and paste sequences from one DNA molecule to another.

9. Sandy B. Primrose and Richard M. Twyman, *Genomics: Applications in Human Biology* (Malden, MA: Blackwell Publishing, 2004), 24.

10. http://www.ncbi.nlm.nih.gov/entrez/dispomim.cgi?id=141900&a=141900_AllelicVariant0243 (accessed November 28, 2008).

11. The International HapMap Project, http://www.hapmap.org/index.html.en (accessed November 28, 2008).

12. Kobilinsky et al., *DNA*, 45–49.

13. Ibid., 45–54.

14. For example, see http://www.dnatestingcentre.com/FTA_cards.htm (accessed December 5, 2008).

15. Bill Hewitt and Vickie Bane, "No Longer Suspects: After 12 Years New DNA Evidence Leads the Boulder D.A. to Apologize to the Ramseys," *People* 70, no. 4 (July 28, 2008): 66.

16. Carole McCartney, "LCN DNA: Proof Beyond Reasonable Doubt," *Nature Reviews Genetics* 9, no. 5 (May 2008): 325.

17. Calculating allele frequencies in populations is beyond the scope of this chapter, but for an easy-to-understand explanation of how to use the Hardy-Weinberg concept to predict allele frequencies, see Ricki Lewis, *Human Genetics: Concepts and Applications*, 8th ed. (Boston: McGraw-Hill, 2008), 269–95.

18. Kobilinsky et al., *DNA*, 47–50.

19. For more information, see "President's DNA Initiative: Advancing Justice through DNA Technology," http://www.dna.gov/uses/database/codis (accessed December 5, 2008).

20. Wayt W. Gibbs, "Shrinking to Enormity: DNA Microarrays Are Reshaping Basic Biology—But Scientists Fear They May Soon Drown in the Data," *Scientific American* 284, no. 2 (February 2001): 33–34.

21. http://www.affymetrix.com/products_services/research_solutions/methods/custom_gene_expression.affx (accessed December 6, 2008).

22. http://www.affymetrix.com/promotions/genetitan1/genetitan.affx (accessed December 6, 2008).

23. https://www.gtldna.net (accessed April 27, 2009).

24. M. R. Cummings, *Human Heredity: Principles and Issues*, 8th ed. (Belmont, CA: Brooks/Cole, 2008), 317.

25. http://www.ncbi.nlm.nih.gov (accessed December 6, 2008).

26. http://www.ornl.gov/sci/techresources/Human_Genome/faq/faqs1.shtml (accessed December 13, 2008).

27. http://www.thegeneticgenealogist.com/2008/10/06/complete-genomics-will-sequence-your-entire-genome-for-5000-starting-in-2009 (accessed December 6, 2008).

28. David A. Wheeler et al., "The Complete Genome of an Individual by Massively Parallel DNA Sequencing," *Nature* 452, no. 17 (April 2008): 872–76.

29. German National Ethics Council, *Biobanks for Research*, ed. Spiros Simitis (Berlin, 2004), 83–86.

30. United Nations Educational Scientific and Cultural Organization, "International Declaration on Human Genetic Data," June 25–27, 2003, http://portal.unesco.org/shs/en/ev.php-URL_ID=2509&URL_DO=DO_TOPIC&URL_SECTION=201.html (accessed December 7, 2008).

31. Alan Petersen, "Biobanks: Challenges for Ethics," *Critical Public Health* 15, no. 4 (December 2005): 303–10.

32. https://www3.nationalgeographic.com/genographic/about.html (accessed December 7, 2008).

33. James Shreeve, "The Greatest Journey Ever Told: The Trail of Our DNA," *National Geographic* (March 2006): 60–73.

34. Family Tree DNA, "Family Tree DNA Y-DNA 12 Test Match 12 for 12," http://www.familyTreeDNA.com (accessed May 19, 2008).

35. Joan Stephenson, "The World of Medicine: Genetic Privacy," *Journal of the American Medical Association* 300, no. 15 (October 15, 2008): 1752.

36. Nils Homer et al., "Resolving Individuals Contributing Trace Amounts of DNA to Highly Complex Mixtures Using High-Density SNP Genotyping Arrays," *PloS Genetics* 4, no. 8 (August 2008), http://www.ncbi. nlm.nih.gov/pubmed/18769715 (accessed November 26, 2008).

37. For further information, see "DNAPrint Genomics," http://www. dnaprint.com/welcome/productsandservices/forensics (accessed April 27, 2009).

38. Amade M'charek, "Contrasts and Comparisons: Three Practices of Forensic Investigation," *Comparative Sociology* 7, no. 3 (August 2008): 387–412. For additional information on DNAWITNESS software, see http://www.bioforensics.com/conference07/Racial_Identification/index. html (accessed December 8, 2008).

39. Timothy Caulfield, Amy L. McGuire, Mildred Cho, Janet A. Buchanan, et al., "Research Ethics Recommendations for Whole-Genome Research: Consensus Statement," *PloS Biology* 6, no. 3 (March 2008): 430–35.

40. Stephen Breyer, "Furthering the Conversation about Science and Society," in *DNA and the Criminal Justice System: The Technology of Justice*, ed. David Lazer (Cambridge, MA: MIT Press, 2004), 13.

41. Ibid., 16.

42. Ibid., 13–22.

PART III

The Future: Where Do We Go from Here?

9

And Away We Go: Opportunities, Challenges, and Social Justice

It's a book of instructions, a record of history,
A medical textbook, it's all these entwined
It's of the people, by the people,
It's for the people, it's yours and it's mine.

—Francis Collins, *Language of God*

INTRODUCTION

We have finally succeeded in mapping the human genome, and this has serious implications for all of us. In particular, it raises issues of accessibility. In this poem, Francis Collins, head of the Human Genome Project, writes about the human genome and reminds us that it belongs to all of us. It is that which all humans share in common, and thus we need to find ways to ensure that the benefits of our science and our technological discoveries helpfully affect us as a whole, and not just privileged populations. All that we have considered up to this point has been a precursor to the bottom-line issue of how this new technology will benefit or harm individuals—in other words, to consider who is likely to benefit and who is likely to not benefit. As one writer observed: "The human genome is now mapped, the Project is now projectile. To whose benefit is it aimed?"[1] It is obvious that there will be some "winners" and some "losers," at least in the short run and possibly for the foreseeable future. We have already addressed some of the ethical issues with regard to specific technologies. But now we explore genetic technologies as a whole through the lens of the concept and principle of justice.

One way that justice is determined is by weighing the burdens and benefits to different groups in society, and to calculate if it is fair. This is more than a utilitarian calculation, in which decisions are typically made in terms of the greatest good for the greatest number. In medicine, some discoveries might benefit just a few in the beginning, but then have the potential to yield benefits for the wider population, especially in the long term. Also, the principle of justice cannot simply dismiss those who will not benefit. We briefly raised some of these justice concerns in past chapters, but we deal with them more specifically here, as well as raise some additional issues.

First, we explain the concept of justice by defining it and by exploring different theories of justice, as well as different kinds of justice. Second, we talk about the groups who are likely to benefit from these technologies, as well as those who are unlikely to benefit from these technologies. Then we discuss some additional ways of exploring justice with regard to genetic technologies. Finally, we offer some concluding comments to the chapter, as well as to the book as a whole.

THE CONCEPT OF JUSTICE

All of us have probably seen at some time the depiction of justice as a blindfolded woman who has a scale in her left hand. The scale represents the idea of weighing evidence for a verdict, and the fact that she is blindfolded means that the decision rendered will be objective. Thus, when it comes to ensuring justice, we must be fair and not discriminate. But to say that justice should be blind does not mean that there are no parameters and guidelines for how we should distribute the goods and services in a society.

What precisely is meant by justice? Justice ultimately has to do with fairness, with "treating equals equally," although this does not mean that in every single case every single person must be treated exactly the same. Justice goes back to the ancient Greeks as one of the virtues, and is also one of the four cardinal virtues in the Christian tradition (the others being prudence, fortitude, and temperance). Justice is also one of the principles of health care (some of the other primary ones being beneficence, nonmaleficence, and autonomy). Social justice in particular has also been considered one of the hallmarks of a good society. In general, justice can be defined as that virtue which enables and requires each person and society to give to others what is their "due." Justice is related to the concept of rights, and is expressed in an external obligatory act toward the other. Within the concept is the idea that there are limits to justice and that some debts cannot be paid in full.[2]

Several different theories about justice exist, even though there may be some generally agreed upon common features. Not everyone agrees, however, that there is a universal conception of justice, because justice is always based on rationality, which must by necessity be the rationality of a particular tradition (and hence historically conditioned).[3] But even those who believe in a universal conception of justice disagree, not so much on what it is but on how we can determine how it gets applied. In her book, *Six Theories of Justice*, Karen Lebacqz provides an excellent overview of three philosophical/liberal and three Christian theories of justice.[4] For the philosophical theories, she focuses on utilitarianism (John Stuart Mill), social contract theory (John Rawls), and entitlement theory (Robert Nozick). For the Christian theories, she focuses on Catholicism (particularly *The Pastoral Letter on Catholic Social Teaching and the U.S. Economy of the National Conference of Catholic Bishops*), Protestantism (Reinhold Niebuhr), and the liberationist tradition (focusing on Jose Porfirio Miranda's *Marx and the Bible*.)

Utilitarianism is the ethical theory that determines the rightness of acts by applying the principle of utility: trying to determine which course of action will bring about the greatest good for the greatest number. The good is generally considered to be happiness, which is understood as the presence of pleasure and the absence of pain (pleasure need not only be understood as physical). Ultimately, then, justice is determined with reference to the principle of utility, and thus justice will be that which applies the principle of utility. One of the ways that John Stuart Mill tried to determine what was just was by identifying six common circumstances that were unjust: depriving people of things to which they have a legal right; depriving people of things to which they have a moral right; people not obtaining what they deserve, breaking faith with people, being partial, and treating people unequally. Utilitarianism's theory of justice is helpful in that it provides concrete guidelines and does put happiness as an important foundation of human life, but it also allows for some to be disadvantaged at the expense of others.

John Rawl's social contract theory, on the other hand, offered a creative solution that is somewhat akin to the blindfolded woman as justice (presented in his book *A Theory of Justice*). He argued for a concept of justice in which decisions about distribution of goods would be made behind what he called a "veil of ignorance." Thus, the parties choosing the principles for the application of justice would lack certain kinds of knowledge that would make the bargaining unfair, such as knowing their place in society and their life plans and goals. They would need to have some knowledge about economic theory, social organization, and human psychology. They would need to be disinterested in the outcome, to be rational, and to be not envious. He believed that these individuals would then choose two principles by which justice would apply: equal liberty and fair equality of opportunity. This theory does allow for some inequities, but they would not exist for all time and individuals could fairly compete for benefits. Two ways that this approach to justice would differ from utilitarianism would be that it would be by rational choice in a fair setting, and would protect the least advantaged. Although Rawls' theory has been subject to a flurry of criticisms, this also attests to its importance as a work with which scholars need to reckon.

The third philosophical theory is based on an analysis of Robert Nozick's important book, *Anarchy, State, and Utopia*. He argued that there needed to be near absolute rights that would provide the foundation of morality, that there would be a minimal state which would allow people freedom to form utopian communities that fit their own vision (he thus argued against a single utopian vision of society), that private property is key, and that justice ultimately consists not in the greatest good for the greatest number, nor in protecting the least advantaged, but in allowing for free exchange. Thus, the fact that some individuals may have more than others can be considered unfortunate, but it is not unfair as long as the rules for free choice in exchange have not been violated.

Let us now explore briefly different Christian conceptions of justice. The Catholic Church has a very well-developed theory of justice based on principles of Catholic social teaching going back at least a century. There are three basic affirmations of Catholic social teaching: the inviolable dignity of the human person, the essentially social nature of human beings, and the belief that the goods of nature and social life are given for all people (the latter is usually referred to as the common good). The Church tends to focus on concrete issues of injustice, particularly economic and political injustice, but has shied away from advocating revolution or violent means to correct situations of injustice. Rather, it has focused on concepts such as a living wage, social solidarity, and political participation as aids to ensuring justice. Ultimately, it has maintained that justice should be judged by the plight of the poor; thus, the poor become the measuring stick for what is just. The Church believes that justice should not be based simply on rationality or a utilitarian calculation, but should be rooted in a faith tradition that responds to a loving and just God who desires justice for all people.

One of the criticisms of the Church's position is that it does not deal seriously enough with sin, which becomes the focus of the Protestant ethicist Reinhold Niebuhr's position. Although he maintained that Christian ethics should begin with love, he also believed that we need to be realistic about sin (his view of ethics has typically been referred to as Christian realism). He believed that the foundation of justice should be freedom and equality. He thought that justice should be established by reason, but since reason too is "fallen," justice also requires the use of power (political and economic) or coercion to establish order. Thus, unlike the Catholic bishops, he did allow for the possibility of violence to establish justice.

Finally, liberationist approaches in Christianity tend to stress the reality of conflict in society, view sin primarily as a structural problem, and emphasize the importance and meaning of history. This focus on justice begins with the perspective of the poor and oppressed, from the reality of the situation of those who are materially and socially disadvantaged and marginalized. Their cry for justice is an attack on the entire system or social order, not simply on individual acts of injustice. Their theory is rooted in Marxist analysis, but is also rooted in the idea of a God who is the liberator of the poor, with the story of Exodus being a key narrative. Because God is the liberator of the poor, a test of our Christianity is that we model ourselves after God and also work to liberate the poor and oppressed, sometimes even to the point of shedding blood.

Despite their sometimes obvious points of emphasis and disagreement with regard to their different conceptions of justice, they all consider justice to be an important feature of a good society, that we should make significant efforts to establish justice, and that fairness with regard to access to good tends to be at the very core of justice.

Another way to explore the concept of justice is by looking at the different kinds of justice that exist; we briefly consider three major ones. Commutative justice generally has to do with issues of justice and injustice between individuals (or between individuals and a group or larger entity). This kind of justice is often represented in courts of law, and can be seen everyday on daytime television shows such as *Judge Judy* and *Judge Joe Brown*. These types of shows very nicely demonstrate examples of commutative justice. Commutative justice deals with situations where there once existed a situation of fairness between parties but because of actions on the part of one party, a situation of unfairness now exists for the other party. One does not even have to go to court for examples of this. Person A asks Person B to loan them $100, and Person B does so. Person B is now down $100. If Person A pays back Person B, however, especially within the specified time period, then they are back to their situation of fairness. But if Person A refuses to pay back Person B for any number of reasons, such as arguing it was a gift and not a loan, or by deciding that it is not important to them to repay this debt, then Person B is in a position of being worse off now than before the transaction. Thus, Person B (and most of us would probably agree) has been treated unfairly by Person A. The only way to really rectify this situation is to somehow require Person A to pay back Person B. This may be an obvious example, but real-life situations, even those represented on these television court shows, are not so easy to adjudicate. For example, Person C's dog accidentally gets loose and bites Person D's daughter, which requires extensive medical bills. Was Person C fully guilty, partly guilty, or not guilty? The answer will ultimately determine what compensation, if any, Person D should get. Person E has Person F (a contractor) do work on her house that she is not happy with, and thus refuses to pay Person F the remaining amount agreed upon for the job. Person F asserts that the job was done according to specifications. Who is right and therefore should be compensated? In all cases of commutative justice, then, efforts should be made to compensate the wronged person by considering all the facts in a case.

A concrete case where commutative justice might come into play with regard to genetic technologies would be situations where a person's DNA was obtained for one purpose and then used for another purpose without their consent, or when a person's DNA is now the property of another. The last is a subject of concern particularly with regard to population genetic studies of indigenous groups, who maintain that companies are benefiting from their DNA without them being compensated.[5] Another example could be where an employer or insurer gains access to an individual's DNA, runs genetic tests on them, and subsequently discriminates against them on the basis of their genetics. Again, many of us would consider this unfair and something for which the individual should be compensated, not to mention it is now illegal given the new GINA legislation.

Retributive justice is another kind of justice; it is generally part of law enforcement and the criminal "justice" system. Retribution tends to do with punishing someone for an offense they committed because we do not think it is "fair" that they get away with committing a crime. Thus, Person M goes into a store and robs the storeowner, Person N, of all of the money he has made for the day, and Person M is subsequently apprehended by the police. Now we do not think that the police should let Person M go free not only because he has committed a crime that has harmed Person N but also because he may prove to be a continuing threat to other storeowners in the community. So, Person M gets some kind of fine or prison sentence, depending on the amount of money taken and why he did it. If it cannot be pleaded out, there could be a court case. But the concept of justice says that there should be some kind of penalty for people who commit crimes because it would not be fair to let them go. In criminal justice, however, distinctions are made among different types of crimes. Thus, Person M gets 1 year in jail for his crime, but we would not consider it to be fair that he receive the death penalty for robbing the storeowner.

Let us look at another situation. Person P kills Person Q on the street. Obviously, we could consider this to be a very serious crime, and thus when Person P is apprehended by the police, a trial is conducted and she is found guilty. She would tend to get a stiffer punishment than the person who committed robbery because we think that a person killing another is more serious than robbing them. Thus, Person P receives a life sentence in a maximum security prison. It would certainly be unfair to give her only 1 year, or even a lesser sentence than the robber received. In addition, criminal law also makes distinctions between the reasons why crimes are committed (motive) and the consequences of the crime. An important issue in the attribution of guilt and the assignment of a sentence would be the reason why Person P did this. We would assess Person P's guilt differently if she did it because she was mentally deranged (insanity or limited cognitive ability), she did it premeditatedly after planning it for 6 months (premeditated murder), she did it in self-defense, she did it in a moment of rage (crime of passion), and so on. In other words, we would distinguish which reason it was and give the sentence accordingly. So, it would be unfair to give the death sentence to someone truly killing in self-defense, and equally unfair to give someone who premeditatedly killed just a stiff fine. Distinctions are also made in criminal justice with regard to results. If Person P shoots Person Q and only injures him although she meant to kill him, this is called attempted murder. But if she actually succeeds in killing him (assuming it was premeditated), it would be called first-degree murder; hence, the sentence would be different for each one. Thus, retributive justice has to do with making sure that society is protected from those who present a danger to it, and punishing them in a degree relative to the crime they committed.

There are a few places where retributive justice could come into play with regard to genetic technologies. One area is with regard to culpability. As was discussed in an earlier chapter, if we find that there is a strong genetic basis for violent behavior, this would affect our assessment of guilt as well as the kind of sentence rendered. Another area would be whether or not there should be a DNA database for law enforcement that would contains samples of everyone's genetic material. Some oppose this as "unfair" because it violates privacy. Others worry that someone could "steal" a sample of their DNA and put it at a crime scene, thereby making them a likely suspect in a crime with which they had nothing to do. Another area where genetics is already affecting retributive justice is where lawyers are arguing that their defendants who have been found guilty of a crime in the past should have their DNA tested against samples found at crime scenes, with the possibility of exonerating them, because of more accurate technologies for DNA analysis than may have existed when their clients originally were convicted.

Distributive justice is probably that area of justice that applies most specifically to genetic technologies. This kind of justice is more about fairness than equality, and has to do with how society can fairly distribute the burdens and benefits, the goods and services, to individuals and groups. It is an especially important concept in health care. Distributive justice has two important components. The first has to do with determining and weighing who is burdened by something and who is benefited. The second has to do with how we can evenly distribute the goods and services of society. Distributive justice is very important when we do not have enough goods to go around and must make tough decisions for how to distribute them. In the next two sections, we explore who we think is likely to benefit from, and who we think is likely to be burdened by, many of the new genetic technologies. But first we more fully describe different possibilities for how goods and services can be distributed in a society by looking at several examples.

There are some things in society that we think it is fair that everyone has access to, and which we can in fact provide everyone access to. One example is free education through the high school level. Because we have enough schools and teachers to go around, we can distribute education to everyone, although it is still possible to argue that the quality of education is not the same for everyone, such as for the rich and the poor. But at least theoretically we can argue that we should provide basic education to everyone. However, for most goods and services in society, we do not have enough for everyone and thus we must set criteria. Different criteria can be set in different circumstances. A good example from health care is that we make available the flu vaccine every year to those who want it and can afford it. Generally we are able to distribute it equally because there is enough for all American citizens who want the flu shot (generally 100 million out of more than 300 million American citizens). But a few years ago we had a shortage and initially only 50 million flu shots were available. Criteria for who should receive it needed to be set by the medical establishment, and this was done on the basis of those who were more likely either to get the flu or to suffer ill effects from it: the elderly, children, those with respiratory problems, and health care workers. Thus, even though we could not treat everyone equally that year, this solution would still be considered "fair" (although ultimately in this case there were enough flu shots for all who wanted it). In fact, "need" is often an important criterion for many areas of health care. For example, those who are most seriously in need of assistance (e.g., heart attack victims) get attention in hospital emergency rooms before those with less serious illnesses or injuries (e.g., a person with a broken finger). The basis on which organs are distributed (there is only one organ available for the approximately 16 people who need it) is also based on need, as well as on being a good match with the donor. However, other goods in society are distributed in different ways, according to different criteria. The criterion of first-come, first-served is used with regard to concert ticket sales. The criterion of who can afford to pay is used with regard to most consumer goods (those who want and can afford to pay $60,000 for a Corvette can do so but we do not say that everyone should have one as a matter of justice). The criterion of merit is used with regard to promotions and college admittance, as well as with many kinds of awards. What is important in the concept of distributive justice, though, is that the criterion that is set for a particular good or service be considered fair by most people, and that once we set the criterion, we do not make exceptions on the basis of other criteria—that is, that we "treat equals equally." Thus, if the criterion for organs is need, we do not bounce someone to the top of the list because they are rich and pay good money for it. But if that did happen, we would consider it "unfair" and hence "unjust." This is not to say, though, that everyone agrees with the particular criterion set for particular goods and services.

How might the concept of distributive justice apply to genetics? It is obvious that some technologies will be available that we cannot give to everyone equally, such as enhancement, genetic therapies, IVF procedures, or even having one's genome screened for various diseases. Many of these procedures are prohibitively expensive and would generally be available only to those who had the means to afford it. It is very difficult to address the specific technologies with regard to the assignment of criteria, so instead we focus on who are likely to be the "winners" and the "losers" with regard to genetic technologies.

THE WINNERS AND THE LOSERS IN GENETIC TECHNOLOGIES

The concept of justice requires that we ask how different individuals and groups will be affected by genetic technologies. In some ways, we all have the potential to be winners. Some scholars recognize the potential for injustices in genetics but maintain that the benefits of these technologies far outweigh the problems.[6] Obviously, any advances that will help medicine improve human health and will help us maintain greater social order through better forensics would overall be of benefit to society. Those in the developed countries are most likely to benefit, because of better health care and financial resources for development of and accessibility to these technologies. The rich and those who will be able to pay for these technologies, especially at the beginning, will be the primary beneficiaries.

Although some genetic screening even today is required for all new-borns, at least in the United States, many of the newer technologies (and even some of the older ones like IVF) are quite expensive. Thus, those who will be able to pay for the services of screening their embryos, having surrogates for their children, paying other women to be egg donors, using genetic therapies (as they become available), or cloning their pets must have the means to do so. And certainly if we get to the point where we can "design" our children in terms of enhancement, the rich will benefit more than others just as they do with regard to other ways of giving their children the best start in life, such as through private schools and tutoring. Of course, their offspring will benefit.

These technologies are a benefit in general to all those who seek greater procreative liberty. One author maintains that although some feminists believe that these technologies make reproduction less oppressive for women (in the sense that biology and reproduction can be separated), she argues that procreative liberty must be placed in the wider context of human flourishing, of which procreation is just a part.[7] Some middle class and poor individuals can benefit financially from offering to others their services of sperm donation, egg donation, and surrogacy. Sperm, costing anywhere from $300 to $3,000, is not as expensive a commodity as are eggs, costing any-where from $4,500 to $50,000. Surrogates can get tens of thousands of dollars for carrying a child, as well as having their medical bills covered during pregnancy.[8] Significant monies are available for research, so researchers will often benefit financially and in status if they make advances in science and develop new technologies, espe-cially those that have the potential to aid the human species in any number of ways. Those who market technological products and serv-ices such as, for example, providing genetic screening, mapping on individual's DNA, and creating opportunities for greater reproductive options, will gain financially if there is a market out there for their products, which there increasingly seems to be. Law enforcement agencies are likely to benefit from better technology with which to both catch the guilty and exonerate the innocent. Having large data-bases of DNA and the forensic capability to determine guilt and innocence can help maintain order and reduce crime in society. Finally, future generations are likely to benefit in some ways from the bold steps taken today in science, especially as technologies become more affordable. For those adhering to theories of justice that emphasize freedom, emerging genetic technologies certainly seem to offer the promise of greater freedom and choices in a number of areas. But other theories of justice emphasize that we need to pay attention as well to those who will not be able to take advantage of this freedom in some way. Thus, we need to consider who the "losers" in our new genetic technological world will be.

If the rich are to benefit, then it seems obvious that the poor will not benefit as much, if at all. Although it is certainly possible for poor and middle class women to sell their eggs and rent their wombs to those willing to pay, this could also be considered exploitation, even if the parties consent to participate. Most significantly, the poor will not be able to afford the technologies available that would provide both themselves and their children greater opportunities. Those who are sick (which can be any of us at any point) and whose DNA is available to employers, insurance agencies, or the government can be subject to discrimination on the basis of their genetic profile. They may not be hired or have their medical bills covered because of preexisting conditions. Indigenous peoples who donated their DNA for population studies are not likely to yield any financial benefit from companies and laboratories that are now able to use their genetic material to yield financial benefits for themselves, such as through patenting, for example.

The disabled may be more discriminated against than they already are for a number of reasons. If genetic screening for disabilities becomes more widespread, the number of disabled will decrease and thereby so will their power. Parents who refuse to have their embryos screened for disabilities or who refuse to destroy their embryos if they are found to have disabilities may be looked down on. The disabled themselves may be viewed as being a particular burden on society, especially if this burden could have been avoided in some way. We can ask how we can welcome those into our community whose presence we are ultimately trying to eliminate. A number of writers in the excellent volume *Theology, Disability and the New Genetics* address some of these specific concerns. An important question raised is, If genetic variation is normal in a species, how can we even determine what a disability is?[9] Disability itself is a social construct, and our notions of personhood and enhancement should inform how technologies are developed and used, not the other way around.[10] The essence of a good life is not necessarily tied in with good health, especially in the Christian tradition.[11] The impetus against disabilities may lead us down the path to bad eugenics, such that we consider individuals with Down syndrome, for example, as having lives not worth living.[12] Finally, it is important to remember that all of us are vulnerable to disease and disability, and thus we are not so different from the disabled.[13]

Embryos and fetuses are likely to be harmed, especially for those who consider them to be persons, humans, or even potential humans or persons. As was discussed in the stem-cell chapter, even those technologies that can possibly avoid harming embryos (and some of them cannot avoid it) had to use embryos along the way. In particular, genetic screening, IVF procedures, and embryonic stem-cell research all have the potential to seriously harm embryos. In addition, all citizens have the potential to be harmed in terms of privacy rights, genetic discrimination, and even criminal justice. With the advent of large DNA databases, individuals may lose control over their genetic material, which could then be put to any number of uses, some of which may have potentially harmful effects. Some have raised concerns about the effect that genetic screening abilities may have even on dead bodies. For example, it is postulated that Abraham Lincoln may have had Marfan's syndrome (which results in abnormally long and gangly limbs), but some scientists suggest that remnants of his DNA, kept in the National Museum of Health and Medicine, be tested.[14] Aside from the fact that Lincoln is dead and cannot consent to this procedure, doing this could be considered a violation of bodily integrity. With more recent deceased individuals, there is the added problem of the effect on their loved ones.

Finally, a group of often overlooked "losers" with regard to genetic technologies are animals. The subject of animal experimentation, and in fact the foundational question of the moral status of animals, is quite controversial. Many excellent works have been written on this subject.[15] The main argument in favor of animal experimentation, particularly in medicine, is that it is likely to yield considerable benefits for humans. Virtually all new procedures and technologies are tested on animals before they can be made available to humans. This is certainly true with regard to genetic science. Let us take the case of Dolly. Although cloning her was an ultimate "success," it also took 277 tries to do this, with many deformed animals along the way. Transgenic animals have been created without regard as to how this might benefit or harm the animal. For example, we can add human hormones to cows that will enable them to produce more milk, but it often creates mastitis and other painful conditions in the cows themselves. The Doogie mouse (so-named after the TV character Doogie Houser, a teenager whose advanced intellectual ability enabled him to be a doctor) was created by scientists enhancing the intellectual abilities of mice, which resulted in them having a much higher level of intelligence than the average mouse. Although initially hailed as a promising breakthrough with the potential for increasing human intelligence, we later found out that the same mechanism that increased intelligence also resulted in considerable pain. It is obvious that whatever one thinks about animal experimentation, genetic science has ensured that animals will continue to be widely used, and not usually in ways that will benefit them.

ADDITIONAL JUSTICE CONSIDERATIONS

Another way of examining the concept of justice from the perspective of emerging genetic technologies is by raising some overall considerations that should be part of the ethical discussion as we move forward. Although the GINA law is an attempt to protect U.S. citizens from genetic discrimination, there is no guarantee that this will happen, and even if it does, it does not eliminate some other potential problem areas, especially globally. A concern particularly within Christian conceptions of justice has to do with what are termed vulnerable populations. This group typically includes the poor, embryos, the disabled, and the elderly[16]—any group that is on the margins of society. As described previously, both the Catholic and the liberationist conceptions begin their analysis from the perspective of the oppressed and marginalized. What would our genetic ethical discussions look like if we did this instead of allowing market forces to set the direction for the ethical debate? We also need to weigh the concerns of developed countries versus developing countries. Because most genetic technologies are designed in developed countries and will be accessible to at least those who can afford it, we can ask the broader question about resources in health care. For example, because funding and resources in medicine are limited, and given the fact that many children and adults die in developing countries from diseases which are easily preventable (e.g., malnutrition), is it legitimate to provide so much money to less significant issues, such as helping individuals choose the sex of their child and genetically modifying their offspring in any number of ways? Of course, it is also possible to argue against this notion that we should defer high-tech enhancements and basic treatments available to all by pointing out that many new, high-tech options start out expensive and eventually become more reasonable.[17] But the question that remains within the Christian tradition is how we can ensure that there truly is a common good. The concept of the common good can certainly be set in contrast to conceptions of justice that emphasize entitlement (Robert Nozick) or the greatest good for the greatest number (utilitarianism). Of course, universal health care for all of the world's citizens could resolve many injustices.

Another important justice concern is the tension between present and future unborn generations. This tension exists in several different areas in ethics, for example, in the environmental legacy that we will be leaving to our children and grandchildren. With regard to genetic technologies, we can ask how our possible use of genetic therapies and modifications (including enhancements) will be passed onto subsequent generations, affecting them in negative ways that we cannot at present see. In fact, we can ask how seriously we even think about future generations with regard to genetic technologies. Some writers have expressed concern that we may in fact be moving toward the world of the "transhuman" or "posthuman," in which the very essence of human nature will be modified. We need to think long and hard about how present technologies will affect future generations, and proceed cautiously where there is likely to be harm.

How much freedom should the market ultimately have to set the genetic technological agenda? Voices come from both sides on this issue; some argue in favor of a liberal market economy with regard to these technologies,[18] and others offer a more cautious perspective. Deborah L. Spar argues that especially with regard to the "baby business," that there is a flourishing market for both children and their component parts, and thus we need to understand this market so that governments can play a more active role in regulating the baby trade.[19]

In addition to the baby trade, there is a proliferation of companies marketing all kinds of genetic technologies, from cloning pets to genetic screening tests to having a picture developed from one's genome. It seems all of these market forces need to at least be monitored in some way. Finally, whose voices should prevail in the continuing discussion of justice implications of genetic technologies? It is virtually a truism of the postmodern world that none of us can be truly objective observers—that we tend to view the world from our own perspective, which includes our educational level, country of origin, financial situation, personal experiences, and so on.

In fact, it has been argued that one's social placement will affect one's view of the HGP. We certainly cannot view the HGP objectively as long as discrimination exists in society in terms of gender, race, class, and nationality.[20] We can see in terms of our discussion of justice, though, that the concerns are not so much with science as with technology. As John Polkinghorne reminds us: "Science gives us knowledge, a gift that is surely always welcome as providing a better basis for decisions than ignorance. But then science's lusty offspring, technology, uses that knowledge to give us power, the ability to do things not previously thought to be possible. This is a more ambiguous gift, since not everything that can be done, should be done."[21]

CONCLUSION

The benefits and burdens of new genetic technologies will be difficult to measure. As we have seen, the same technology, such as donating eggs, can be a benefit and a burden—even to the same person! When we get into the arena of who will have access, this is a hard area in which to come up with equitable solutions. There is the additional problem of different conceptions of justice, and we are not sure which understanding should prevail. But despite these difficulties, most philosophers and theologians agree that justice ultimately has to do with fairness, even if how we get there or what the results might be are understood in diverse ways. We think that justice is and should be at the heart of ethical discussions related to genetic science and technologies. It also should not simply be an afterthought or an add-on; it should be at the forefront as these technologies are developed, and perhaps even before they are made available to the public. In fact, although many writers focus significant attention on some of the ethical issues raised by these emerging genetic technologies, relatively scant attention has been paid to the concept of justice and these technologies. This should prove to be a fruitful arena for further work in the future.

As the previous chapters indicate, important ethical issues must be considered regarding emerging genetic technologies. Because so many ethical questions are raised by the new genetics, one author has suggested that we develop a whole new field to deal with ethical issues in genetics. Thus, instead of viewing it as a simple subset of medical ethics or bioethics, it would be a field of study in itself—she terms it genomorality—that would include public policy discussion, especially about social justice and genetic technologies.[22] Whether or not this is done, it is obvious that much work in ethics and genetics must be done. The fact that the ELSI Project is alive and well we consider to be a very good thing indeed.

In this book, we have traversed over a wide territory. We have tried to present a truly interdisciplinary approach by exploring the science behind the technologies as well as the ethical issues that arise because of the development of these technologies. We do believe that understanding the science behind these issues helps inform our ethics in a way that it cannot fully do without our understanding the science. We have explored the religion and science relationship, particularly biology and Christian theology, and concluded that each discipline has much to learn from the other. We have provided some background on the HGP and the ELSI Project. We have addressed specific issues with regard to the science behind the technologies and ethical concerns, such as with designing our children, stem-cell research, cloning, genes and behavior, and genes and privacy. We have focused on justice concerns because we think that they need to be at the heart of the ethical discussion of these new technologies.

We want to offer some final comments with regard to what we think needs to happen in the present and especially for the future. We believe that a continuing dialogue is necessary among many disciplines, not just between biology and theology. The more voices that there are in the discussion, the more likely we will be able to target a wide range of issues and concerns that might get overlooked by looking at genetic technologies simply from the lens of one or two disciplines. A statement made by one of the founders of utilitarianism, John Stuart Mill (in *On Liberty*, 1859), put it very nicely: "Because he has felt, that the only way in which a human being can make some approach to knowing the whole of a subject, is by hearing what can be said about it by persons of every variety of opinion, and studying all modes in which it can be looked at by every character of mind. No wise man ever acquired his wisdom in any mode but this; nor is it in the nature of human intellect to become wise in any other manner." We whole-heartedly agree! We also want to emphasize that no matter what limits the United States puts on genetic research, it will be done elsewhere. Thus, what we need to ensure is that discussion of ethical implications continues, as well as attempts to develop mechanisms so that regulations and regulatory agencies are put in place to anticipate concerns and address current and emerging genetic technologies.

We realize the limitations of a book such as this. With the plethora of materials coming out on the new genetics, we simply could not address everything, including some very important issues such as the perspective of other world religions, or even more detail on some of the current technologies. But we did try to give an up-to-date overview of what is currently available as well as the ethical and religious discussion on these issues. We know that even as soon as the book has gone to press, there will be new techniques and technologies. We know, though, that even if there are new technological advances and new discoveries, many of the ethical and religious questions will remain the same. We hope that you now have some tools to help you consider those questions as well as to understand some of the basic science that drives both the technology and the resulting ethical questions. It is our hope that you will keep yourselves apprised of new developments and continuing discussion in the area of genetics; these emerging genetic technologies are not going away, and they have the ability to affect all of our lives in significant ways.

NOTES

1. Peter Scott, "Is the Goodness of God Good Enough? The Human Genome Project in Theological and Political Perspective," in *Brave New World? Theology, Ethics, and the Human Genome*, ed. Celia Deane-Drummond (London and New York: T&T Clark, 2003), 314.

2. Josef Pieper, *The Four Cardinal Virtues* (Notre Dame, IN: University of Notre Dame Press, 1965), 43–63, 104. This is an excellent little volume on these four virtues.

3. Alasdair MacIntyre makes this argument in his well-known book, *Whose Justice? Which Rationality?* (Notre Dame, IN: University of Notre Dame Press, 1988).

4. The summary that follows has come from her book, *Six Theories of Justice: Perspectives from Philosophical and Theological Ethics* (Minneapolis: Augsburg, 1986).

5. Indigenous Peoples Council on Biocolonialism, http://www.ipcb.org (accessed December 21, 2008).

6. For example, Ronald Green, in *Babies by Design: The Ethics of Genetic Choice* (New Haven, CT, and London: Yale University Press, 2007), although he raises the possibility of what he calls "genobility," still thinks that genetic technology is overall a positive and should be encouraged.

7. Karen Peterson-Iyer, *Designer Children: Reconciling Genetic Technology, Feminism, and Christian Faith.* (Cleveland: Pilgrim Press, 2004).

8. For an excellent overview on reproductive services, see Debora L. Spar, *The Baby Business: How Money, Science, and Politics Drive the Commerce of Competition* (Boston: Harvard Business School Press, 2006).

9. John Swinton, "Introduction: Re-imagining Genetics and Disability," in *Theology, Disability, and the New Genetics: Why Science Needs the Church*, ed. John Swinton and Brian Brock (London and New York: T&T Clark, 2007), 3.

10. Brent Waters, "Disability and the Quest for Perfection: A Moral and Theological Inquiry," in *Theology, Disability, and the New Genetics: Why Science Needs the Church*, ed. John Swinton and Brian Brock (London and New York: T&T Clark, 2007), 209.

11. Martina Holder-Franz, "Life as Being in Relationship: Moving beyond a Deficiency-Orientated View of Human Life," in *Theology, Disability, and the New Genetics: Why Science Needs the Church*, ed. John Swinton and Brian Brock (London and New York: T&T Clark, 2007), 64.

12. Mary B. Mahowald, "Aren't We All Eugenicists Anyway?" in *Theology, Disability, and the New Genetics: Why Science Needs the Church*, ed. John Swinton and Brian Brock (London and New York: T&T Clark, 2007), 106.

13. This is the main thesis of Alasdair MacIntrye, *Dependent Rational Animals: Why Human Beings Need the Virtues* (Chicago: Open Court, 1999).

14. Claudia Kalb, "The Secrets in Lincoln's DNA," *Newsweek*, February 23, 2009.

15. Because we could not go into a lot of detail here on this subject, we refer you to Donna Yarri, *The Ethics of Animal Experimentation: A Critical Analysis and Constructive Christian Proposal*, AAR Academy Series (Oxford: Oxford University Press, 2005); the bibliography of this book contains many helpful works.

16. Cynthia Y. Read, Robert C. Green, and Michael A. Snyder, eds., *Aging, Biotechnology, and the Future* (Baltimore: Johns Hopkins University Press, 2008). These authors have compiled an excellent collection of essays in which they look in particular at how genetic technologies will affect the elderly, particularly with regard to social support.

17. John Harris, *Enhancing Evolution: The Ethical Case for Making Better People* (Princeton, NJ: Princeton University Press, 2007), 32.

18. See in particular Green, *Babies by Design*, and Nicholas Agar, *Liberal Eugenics: In Defence of Human Enhancement* (Malden, MA, and Oxford: Blackwell Publishing, 2004).

19. Spar, *The Baby Business*, xv–xviii.

20. Scott, "Is the Goodness of God Good Enough?"

21. John Polkinghorne, *Exploring Reality: The Intertwining of Science and Religion* (New Haven, CT, and London: Yale University Press, 2005), 147.

22. Julie Clague, "Beyond Beneficence: The Emergence of Genomorality and the Common Good," in *Brave New World? Theology, Ethics, and the Human Genome*, ed. Celia Deane-Drummond (London and New York: T&T Clark, 2003).

WORKS CITED

Advanced Cell Technology. http://www.advancedcell.com.

Affymetrix, Inc. http://www.affymetrix.com/index.affx.

Agar, Nicholas. *Liberal Eugenics: In Defence of Human Enhancement*. Malden, MA, and Oxford: Blackwell Publishing, 2004.

American Association for the Advancement of Science. "Statement on Human Cloning." February 14, 2002. http://www.aaas.org/news/releases/2002/Cloning.shtml.

BabyHopes. http://www.babyhopes.com.

Barbour, Ian G. *Ethics in an Age of Technology*. Gifford Lectures, vol. 2. San Francisco: Harper San Francisco, 1993.

———. *Nature, Human Nature, and God*. Minneapolis: Fortress Press, 2002.

———. "On Typologies for Relating Science and Religion." *Zygon: Journal of Religion and Science* 37, no. 2 (June 2002): 345–59.

———. *Religion in an Age of Science*. Gifford Lectures, vol. 1. San Francisco: Harper Collins, 1990.

Barnard, Jeff. "A Scientist Reaches Out to Evangelicals in His New Book Expressing Concern for the Health of the Environment." *Reading Eagle*. November 18, 2006, A9.

Barron, James. "Biotech Company to Auction Chances to Clone a Dog." *New York Times*. May 21, 2008.

Baruch, Susannah, and Kathy Hudson. "Civilian and Military Genetics: Nondiscrimination Policy in a Post-GINA World." *American Journal of Human Genetics* 83, no. 4 (October 10, 2008): 435–44.

Battey, James F., Chair, NIH Stem Cell Task Force, and Director, National Institute on Deafness and Other Communication Disorders. "Senate Testimony: Alternative Methods of Obtaining Embryonic Stem Cells." In *Testimony before the Subcommittee on Labor, Health and Human Services, Education, and Related Agencies*. Washington, DC: NIH, July 12, 2005.

Bay, Michael. *The Island*. DreamWorks LLC and Warner Brothers Entertainment, Inc., 2005.

BBC News. "Details of Hybrid Clone Revealed." June 18, 1999. Sci/Tech. http://news.bbc.co.uk/2/hi/science/nature/371378.stm.

———. "Public Debate on Hybrid Embryos." January 11, 2007. Sci/Tech. http://news.bbc.co.uk/2/hi/health/6251627.stm.

Beauchamp, Tom L., and James F. Childress. *Principles of Biomedical Ethics*. 5th ed. New York and Oxford: Oxford University Press, 2008.

Beckwith, Jonathan. "Whither Behavioral Genetics?" In *Wrestling with Behavioral Genetics: Science, Ethics, and Public Conversation*, ed. Eric Parens, Audrey R. Chapman, and Nancy Press. Baltimore: Johns Hopkins University Press, 2006.

BioCentre. Council of Europe and the United Nations. http://www.bioethics.ac.uk/index.php?do=topic&sid=13.

Biopolis. One-North Development Group, JTC Corporation. http://www.one-north.sg/hubs_biopolis.aspx.

Blaese, R. Michael. "Germ-Line Modification in Clinical Medicine: Is There a Case for Intentional or Unintended Germ-Line Changes?" In *Designing Our Descendants: The Promises and Perils of Genetic Modification*, ed. Audrey R. Chapman and Mark S. Frankel, 68–76. Baltimore: Johns Hopkins University Press, 2003.

Bluestone, Mimi. "Science and Ethics: The Double Life of Nancy Wexler." *Ms.: The World of Women* 2, no. 3 (November/December 1991): 90–91.

Bottkin, Jeffrey R., William M. McMahon, and Leslie Pickering Francis, eds. *Genetics and Criminality: The Potential Misuse of Scientific Information in Court*. Washington, DC: American Psychological Association, 1999.

Bouma, Hessel, III. "Completing the Human Genome Project: The End Is Just the Beginning." *Perspectives in Science and Christian Faith* 52, no. 3 (2000): 152–55.

Breyer, Stephen. "Furthering the Conversation About Science and Society." In *DNA and the Criminal Justice System: The Technology of Justice*, ed. David Lazer, 13–22. Cambridge, MA: MIT Press, 2004.

Bronowski, Jacob. *The Ascent of Man*. Boston and Toronto: Little, Brown, 1973.

Brooke, John Hedley. *Science and Religion: Some Historical Perspectives*. Cambridge and New York: Cambridge University Press, 1991.

Brun, Rudolf B. "Cloning Humans? Current Science, Current Views, and a Perspective from Christianity." *Differentiation* 69 (2002): 184–87.

California Institute for Regenerative Medicine. 2007. California State Government. http://www.cirm.ca.gov.

Cantor, Geoffrey, and Chris Kenny. "Barbour's Fourfold Way: Problems with His Taxonomy of Science-Religion Relationships." *Zygon: Journal of Religion and Science* 36 (December 2001): 765–81.

Catechism of the Catholic Church. New York and London: An Image Book/Doubleday, 1995.

Caulfield, Timothy, et al. "Research Ethics Recommendations for Whole-Genome Research: Consensus Statement." *PloS Biology* 6, no. 3 (March 2008): 430–35.

Center for Genetics and Society. "U.S. Federal Policies." http://www.geneticsandsociety.org/article.php?id=305.

Chapman, Audrey R. *Unprecedented Choices: Religious Ethics at the Frontiers of Genetic Science*. Theology and Science. Minneapolis: Fortress Press, 1999.

Chung, Young, et al. "Embryonic and Extraembryonic Stem Cell Lines Derived from Single Mouse Blastomeres." *Nature* 439 (January 12, 2006): 216–19.

A Civic Project to Track Congress. http://www.govtrack.us.

Clague, Julie. "Beyond Beneficence: The Emergence of Genomorality and the Common Good." In *Brave New World? Theology, Ethics, and the Human Genome*, ed. Celia Deane-Drummond. London and New York: T&T Clark, 2003.

Clark, Stephen R. L. *Biology and Christian Ethics*. Cambridge: Cambridge University Press, 2000.

Cole-Turner, Ronald. "Principles and Politics: Beyond the Impasse over the Embryo." In *God and the Embryo: Religious Voices on Stem Cells and Cloning*, ed. Brent Waters and Ronald Cole-Turner, 88–97. Washington, DC: Georgetown University Press, 2003.

Cole-Turner, Ronald, ed. *Design and Destiny: Jewish and Christian Perspectives on Human Germline Modification*. Cambridge, MA, and London: MIT Press, 2008.

Collins, Francis S. *The Language of God: A Scientist Presents Evidence for Belief*. New York: Free Press, 2006.

Connors, Russell B., and Patrick T. McCormick. *Character, Choices, and Community: The Three Faces of Christian Ethics*. New York and Mahwah, NJ: Paulist Press, 1998.

Culver, C. M. "Concept of Genetic Malady." In *Morality and the New Genetics*, ed. B. Gert, 147–66. Boston: Jones & Bartlett, 1996.

Cummings, Michael R. *Human Heredity: Principles and Issues*. Belmont, CA: Brooks/Cole, 2009.

Dawkins, Richard. *A Devil's Chaplain: Reflections on Hope, Lies, Science, and Love*. New York: Mariner Books, 2004.

———. *The Selfish Gene*. London: Oxford University Press, 1976.

Deane-Drummond, Celia. "Future Perfect? God, the Transhuman Future, and the Quest for Immortality." In *Future Perfect? God, Medicine, and Human Identity*, ed. Celia Deane-Drummond and Peter Manley Scott, 168–82. London and New York: T&T Clark International, 2006.

Deane-Drummond, Celia, ed. *Brave New World? Theology, Ethics, and the Human Genome*. London and New York: T&T Clark, 2003.

Deane-Drummond, Celia, and Peter Manley Scott, eds. *Future Perfect? God, Medicine, and Human Identity*. London and New York: T&T Clark International, 2006.

DeCODEme. DeCODE Genetics. http://www.decodeme.com.

Deech, Ruth. "30 Years: From IVF to Stem Cells." *Nature* 454, no. 7202 (July 17, 2008): 280–81.

Dimos, John T., et al. "Induced Pluripotent Stem Cells Generated from Patients with ALS Can Be Differentiated into Motor Neurons." *Science* 321, no. 5893 (August 29, 2008): 1218–21.

DNA 11. http://www.dna11.com/about.asp.

DNA Witness, DNAPrint Genomics. http://www.dnaprint.com/welcome/productsandservices/forensics.

Dresser, R. "Genetic Modification of Pre-Implantation Embryos: Toward Adequate Human Research Policies." *Milbank Quarterly* 82, no. 1 (2004): 195–214.

Edwards, F. N., F. N. Schrick, M. D. McCracken, S. R. van Amstel, F. M. Hopkins, and C. J. Davis. "Cloning Adult Farm Animals: A Review of the Possibilities and Problems Associated with Somatic Cell Nuclear Transfer." *American Journal of Reproductive Immunology* 50 (2003): 113–23.

Ellul, Jacques. *The Technological Society*, trans. John Wilkinson. New York: Alfred A. Knopf, 1964.

Extend Fertility, Inc. http://www.extendfertility.com.

Family DNA Tree. Genealogy by Genetics, Ltd. http://www.familytreedna.com.

Fertility Institutes. 2001–2008. http://www.gender-selection.com.

Flaman, Paul. *Genetic Engineering, Christian Values, and Catholic Teaching*. New York and Mahwah, NJ: Paulist Press, 2002.

Fletcher, Joseph. *Situation Ethics: The New Morality*. Philadelphia: Westminster Press, 1966.

Frankena, William K. *Ethics*. 2nd ed. Englewood Cliffs, NJ: Prentice-Hall, 1973.

Genetic Genealogist, The. http://www.thegeneticgenealogist.com.

Genetic News Network. http://www.genomenewsnetwork.org/resources/sequenced_genomes/genome_guide_p1.shtml.

Genetic Privacy. Electronic Privacy Information Center. http://epic.org/privacy/genetic.

Genetics and Criminality: The Potential Misuse of Scientific Information in Court. Washington, DC: American Psychological Association, 1999.

Genetic Testing Laboratories, Inc. https://www.gtldna.net.

Genographic Project, National Geographic and IBM. https://genographic.nationalgeographic.com/genographic/about.html.

German National Ethics Council. *Biobanks for Research*, ed. Spiros Simitis. Berlin: Nationaler Ethikrat, 2004.

Gert, B., ed. *Morality and the New Genetics*. Boston: Jones & Bartlett, 1996.

Gibbs, Wayt W. "Shrinking to Enormity: DNA Microarrays Are Reshaping Basic Biology—But Scientists Fear They May Soon Drown in the Data." *Scientific American* 284, no. 2 (February 2001): 33–34.

Gillon, Raanan, ed. *Principles of Health Care Ethics*. Chichester, UK, and New York: John Wiley & Sons, 1994.

Gilman, Charlotte Perkins. *Herland*. Mineola, NY: Dover, 1998.

Görman, Ulf. "Never Too Late to Live a Little Longer? The Quest for Extended Life and Immortality—Some Ethical Considerations." In *Future Perfect? God, Medicine, and Human Identity*, ed. Celia Deane-Drummond and Peter Manley Scott, 143–54. London and New York: T&T Clark International, 2006.

Gould, James L., and Carol Grant Gould. *The Animal Mind*. New York: Scientific American Library, 1994.

Green, Ronald M. *Babies by Design: The Ethics of Genetic Choice*. New Haven, CT, and London: Yale University Press, 2007.

———. "Can We Develop Ethically Universal Embryonic Stem-Cell Lines?" *Nature Reviews Genetics* 8 (June 2007): 480–85.

———. *The Human Embryo Research Debates: Bioethics in the Vortex of Controversy*. Oxford and New York: Oxford University Press, 2001.

———. *Religion and Moral Reason: A New Method for Comparative Study*. New York and Oxford: Oxford University Press, 1988.

———. *Religious Reason: The Rational and Moral Basis of Religious Belief*. New York: Oxford University Press, 1978.

Habermas, Jürgen. *The Future of Human Nature*. Cambridge: Polity Press, 2003.

Hamer, Dean. *The God Gene: How Faith Is Hardwired into Our Genes*. New York: Anchor Books, 2004.

Harris, John. *Enhancing Evolution: The Ethical Case for Making Better People*. Princeton, NJ: Princeton University Press, 2007.

Hauerwas, Stanley. *A Community of Character: Toward a Constructive Christian Ethic*. Notre Dame, IN, and London: University of Notre Dame Press, 1981.

Haught, John F. *Science and Religion: From Conflict to Conversation*. Mahwah, NJ: Paulist Press, 1995.

Hefner, Philip. *Technology and Human Becoming*. Minneapolis: Fortress Press, 2003.

Hewitt, Bill, and Vickie Bane. "No Longer Suspects: After 12 Years New DNA Evidence Leads the Boulder D.A. to Apologize to the Ramseys." *People* 70, no. 4 (July 28, 2008): 66.

Holder-Franz, Martina. "Life as Being in Relationship: Moving beyond a Deficiency-Orientated View of Human Life." In *Theology, Disability, and the New Genetics: Why Science Needs the Church*, ed. John Swinton and Brian Brock, 57–66. London and New York: T&T Clark, 2007.

Homer, Nils, et al. "Resolving Individuals Contributing Trace Amounts of DNA to Highly Complex Mixtures Using High-Density SNP Genotyping Arrays." *PloS Genetics* 4, no. 8 (August 2008). http://www.ncbi.nlm.nih.gov/pubmed/18769715.

Hudson, Kathy L., M. K. Holohan, and Francis S. Collins. "Keeping Pace with the Times: The Genetic Information Nondiscrimination Act of 2008." *New England Journal of Medicine* 358, no. 25 (19 June 2008): 2661–63.

Human Genome Organisation (HUGO). National Human Genome Research Institute. http://www.hugo-international.org.

Human Genome Project. "Cloning Fact Sheet." September 19, 2008. U.S. Department of Energy, Office of Science, Office of Biological and Environmental Research. http://www.ornl.gov/sci/techresources/Human_Genome/elsi/cloning.shtml.

IJzendoorn, M. H., M. J. Bakermans-Kranenburg, and J. Mesman. "Dopamine System Genes Associated with Parenting in the Context of Daily Hassles." *Genes, Brain and Behavior* 7 (2008): 403–10.

Illmensee, Karl, Khalied Kaskar, and Panayiotis M. Zavos. "In Vitro Blastocyst Development from Serially Split Mouse Embryos and Future Implications for Human Assisted Reproductive Technologies." *Fertility and Sterility* 86 (October 2006): 1112–20.

Indigenous Peoples Council on Biocolonialism. http://www.ipcb.org.

Ingender. 2008. In-Gender.Com. http://www.in-gender.com.

International HapMap Project. http://www.hapmap.org/index.html.en.

Kass, Leon R. *Human Cloning and Human Dignity: The Report of the President's Council on Bioethics.* New York: Public Affairs Books, 2002.

Keefer, C. L., L. Blomberg, and L. Talbot. "Challenges and Prospects for the Establishment of Embryonic Stem Cell Lines of Domesticated Ungulates." *Animal Reproduction Sciences* 98 (October 2006): 147–68.

Kennedy, T. G. "Physiology of Implantation." 10th World Congress on in vitro Fertilization and Assisted Reproduction, May 1997. http://publish.uwo.ca/~kennedyt/t108.pdf.

Keverne, Eric B., and James P. Curley. "Epigenetics, Brain Evolution, and Behavior." *Frontiers in Neuroendocrinology* 29, no. 3 (June 2008): 398–412.

Kim, Kitai, et al. "Histocompatible Embryonic Stem Cells by Parthenogenesis." *Science* 315, no. 5811 (January 26, 2007): 482–86.

Klimanskaya, Irina, et al. "Human Embryonic Stem Cell Lines Derived from Single Blastomeres." *Nature* 444 (November 23, 2006): 481–85.

Kobilinsky, Lawrence, Thomas F. Liotti, and Jamel Oeser-Sweat. *DNA: Forensic and Legal Applications.* Hoboken, NJ: John Wiley & Sons, 2005.

Lebacqz, Karen. *Six Theories of Justice: Perspectives from Philosophical and Theological Ethics.* Minneapolis: Augsburg Publishing House, 1986.

Levine, Aaron D. *Cloning: A Beginner's Guide.* Oxford: Oneworld, 2007.

Lewis, Ricki. *Human Genetics: Concepts and Applications.* 7th ed. Boston: McGraw-Hill, 2007.

———. *Human Genetics: Concepts and Applications.* 8th ed. New York: McGraw-Hill, 2008.

"Love Longears." American Donkey and Mule Society. http://www.lovelongears.com.

Loy, Heather A. "Is Our Fate in Our Genes? Behavior Genetics." In *Science and the Soul: Christian Faith and Psychological Development,* ed. Scott W. Vanderstoep, 151–73. Lanham, MD, and Boulder, CO: University Press of America, 2003.

MacIntyre, Alasdair. *Dependent Rational Animals: Why Human Beings Need the Virtues.* Chicago: Open Court, 1999.

———. *Whose Justice? Which Rationality?* Notre Dame, IN: University of Notre Dame Press, 1988.

Mahowald, Mary B. "Aren't We All Eugenicists Anyway?" In *Theology, Disability, and the New Genetics: Why Science Needs the Church,* ed. John Swinton and Brian Brock, 96–113. London and New York: T&T Clark, 2007.

MayoClinic.com. "Tools for Healthier Lives." 1998–2008. Mayo Foundation for Education and Medical Research. http://www.mayoclinic.com.

McCartney, Carole. "LCN DNA: Proof beyond Reasonable Doubt." *Nature Reviews Genetics* 9, no. 5 (May 2008): 325.

McDaniel, Jay. *Of God and Pelicans: A Theology of Reverence for Life.* Louisville, KY: Westminster/John Knox Press, 1989.

McGrath, Alister E. *Christian Theology: An Introduction.* 3rd ed. Malden, MA, and Oxford: Blackwell Publishers, 2001.

M'charek, Amade. "Contrasts and Comparisons: Three Practices of Forensic Investigation." *Comparative Sociology* 7, no. 3 (August 2008): 387–412.

Merriam-Webster On-Line Dictionary. 2005. Merriam-Webster, Inc. http://www.merriam-webster.com.

Messer, Neil G. "The Human Genome Project, Health, and the 'Tyranny of Normality.'" In *Brave New World: Theology, Ethics, and the Human Genome,* ed. Celia Deane-Drummond, 91–115. London and New York: T&T Clark, 2003.

MicroSort. 2001–2006. Genetics and IVF Institute. http://www.microsort.net.

Midgley, Mary. *Science as Salvation: A Modern Myth and Its Meaning.* London and New York: Routledge, 1992.

Morgan, Rose M. *The Genetics Revolution: History, Fears, and Future of a Life-Altering Science.* Westport, CT, and London: Greenwood Press, 2006.

Morris, John F., ed. *Medicine, Health Care, and Ethics: Catholic Voices.* Washington, DC: Catholic University Press of America, 2007.

Mothers 35 Plus. http://www.mothers35plus.co.uk.

Munafo, Marcus R., and Elaine C. Johnstone. "Genes and Cigarette Smoking." *Addiction* (2008): 1–12.

Mundy, Liza. "A World of Their Own." *Washington Post,* magazine section, March 22, 2002.

National Center for Biotechnology Information, National Library of Medicine and National Institutes of Health. http://www.ncbi.nlm.nih.gov.

National Conference of State Legislatures. "State Human Cloning Laws." http://www.ncsl.org/programs/health/Genetics/rt-shcl.htm.

National Human Genome Research Institute, National Institutes of Health. http://www.genome.gov.

National Society of Genetic Counselors. http://www.nsgc.org.

Navigenics, Inc. http://www.navigenics.com.

Niccol, Andrew. *Gattaca.* Columbia TriStar Pictures, 1997.

Oak Ridge National Laboratory. U.S. Department of Energy. http://www.ornl.gov.

Office of Biotechnology Activities, Office of Science and Policy, National Institutes of Health, Energy Research. http://oba.od.nih.gov/oba/index.html.

Online Mendelian Inheritance in Man (OMIM). National Center for Biotechnology Information (NCBI). http://www.ncbi.nlm.nih.gov/omim.

Parens, Erik, ed. *Enhancing Human Traits: Ethical and Social Implications.* Washington, DC: Georgetown University Press, 1998.

Parens, Erik, Audrey R. Chapman, and Nancy Press, eds. *Wrestling with Behavioral Genetics: Science, Ethics, and Public Conversation.* Baltimore: Johns Hopkins University Press, 2006.

Pathak, Bhavani G. "Scientific Methodologies to Facilitate Inheritable Genetic Modifications in Humans." In *Designing Our Descendants: The Promises and Perils of Genetic Modification,* ed. Audrey R. Chapman and Mark S. Frankel, 55–67. Baltimore: Johns Hopkins University Press, 2003.

PBS Nature. http://www.pbs.org/wnet/nature/episodes/flight-school/the-man-who-walked-with-geese/2656.

Peters, Ted. *Playing God? Genetic Determinism and Human Freedom.* New York and London: Routledge, 2003.

———. "Science and Theology: Toward Consonance." In *Science and Theology: The New Consonance,* ed. Ted Peters, 11–39. Boulder, CO, and Oxford: Westview Press, 1998.

———. *The Stem Cell Debate.* Minneapolis: Fortress Press, 2007.

Peters, Ted, ed. *Science and Theology: The New Consonance*. Boulder, CO, and Oxford: Westview Press, 1998.

Petersen, Alan. "Biobanks: Challenges for Ethics." *Critical Public Health* 15, no. 4 (December 2005): 303–10.

Peterson-Iyer, Karen. *Designer Children: Reconciling Genetic Technology, Feminism, and Christian Faith*. Cleveland: Pilgrim Press, 2004.

Picoult, Jodi. *My Sister's Keeper*. New York: Washington Square Press, 2004.

Pieper, Josef. *The Four Cardinal Virtues*. Notre Dame, IN: University of Notre Dame Press, 1965.

Polkinghorne, John. *Exploring Reality: The Intertwining of Science and Religion*. New Haven, CT, and London: Yale University Press, 2005.

———. *Science and Theology: An Introduction*. Minneapolis: Fortress Press, 1998.

Pontifical Council for Pastoral Assistance. *Charter for Health Care Workers*. Boston: St. Paul Media & Books, 1999.

President's DNA Initiative. U.S. Department of Justice. http://www.dna.gov/uses/database/codis.

Primrose, Sandy B., and Richard M. Twyman. *Genomics: Applications in Human Biology*. Malden, MA: Blackwell Publishing, 2004.

Rachels, James. *The Elements of Moral Philosophy*. 3rd ed. Boston and Burr Ridge, IL: McGraw-Hill College, 1999.

Ramsey, Paul. *Fabricated Man: The Ethics of Genetic Control*. New Haven, CT, and London: Yale University Press, 1970.

Read, Cynthia, Robert C. Green, and Michael A. Snyder, eds. *Aging, Biotechnology, and the Future*. Baltimore: Johns Hopkins University Press, 2008.

Reebs, Stephan. "Time to Split." *Natural History* 117, no. 5 (June 2008): 12.

Resnik, David B. *The Price of Truth: How Money Affects the Norms of Science*. New York: Oxford University Press, 2007.

Ridley, Matt. *Nature via Nurture: Genes, Experience, and What Makes Us Human*. New York: HarperCollins, 2003.

Robinson, Daniel N., Gladys M. Sweeney, and Richard Gill, eds. *Human Nature in Its Wholeness: A Roman Catholic Perspective*. Washington, DC: Catholic University of America Press, 2006.

Rose, Steven. *Lifelines: Biology beyond Determinism*. New York: Oxford University Press, 1998.

Ross, W. D. *The Right and the Good*. Indianapolis, IN, and Cambridge: Hackett Publishing, 1930.

Sandel, Michael J. *The Case against Perfection: Ethics in the Age of Genetic Engineering*. Cambridge, MA, and London: Belknap Press of Harvard University Press, 2007.

Schaffner, Franklin J. *The Boys from Brazil*. Twentieth Century Fox, 1978.

Schaffner, Kenneth F. "Behavior: Its Nature and Nurture, Part I and Part II." In *Wrestling with Behavioral Genetics: Science, Ethics, and Public Conversation*, ed. Erik Parens, Audrey R. Chapman, and Nancy Press, 3–39. Baltimore: Johns Hopkins University Press, 2006.

Scott, Peter. "Is the Goodness of God Enough? The Human Genome Project in Theological and Political Perspective." In *Brave New World? Theology, Ethics, and the Human Genome*, ed. Celia Deane-Drummond, 294–318. London and New York: T&T Clark, 2003.

Shannon, Thomas A. "The Roman Catholic Magisterium and Genetic Research: An Overview and Evaluation." In *Design and Destiny: Jewish and Christian Perspectives on Human Germline Modification*, ed. Ronald Cole-Turner, 55–71. Cambridge, MA, and London: MIT Press, 2008.

Shannon, Thomas A., and James J. Walter, eds. *The New Genetic Medicine: Theological and Ethical Reflections*. Lanham, MD, and Boulder, CO: Rowman and Littlefield, 2003.

Shreeve, James. "The Greatest Journey Ever Told: The Trail of Our DNA." *National Geographic* (March 2006): 60–73.

Spar, Debora L. *The Baby Business: How Money, Science, and Politics Drive the Commerce of Competition*. Boston: Harvard Business School Press, 2006.

Spemann, Hans. *Embryonic Development and Induction*. New Haven, CT: Yale University Press, 1938.

Stephenson, Joan. "The World of Medicine: Genetic Privacy." *Journal of the American Medical Association* 300, no. 15 (15 October 2008): 1752.

Stevenson, Leslie. *Seven Theories of Human Nature*. 2nd ed. New York and Oxford: Oxford University Press, 1987.

Stroll, Avrum. *Did My Genes Make Me Do It? and Other Philosophical Dilemmas*. Oxford: Oneworld, 2004.

Swinton, John. "Introduction: Re-imagining Genetics and Disability." In *Theology, Disability, and the New Genetics: Why Science Needs the Church*, ed. John Swinton and Brian Brock, 1–25. London and New York: T&T Clark, 2007.

Swinton, John, and Brian Brock, eds. *Theology, Disability, and the New Genetics: Why Science Needs the Church*. London and New York: T&T Clark, 2007.

Tech Museum of Innovation: Understanding Genetics. Stanford School of Medicine. http://www.thetech.org/genetics/ask.php?id=86.

United Mitochondrial Disease Foundation. http://www.umdf.org/site/c.dnJEKLNqFoG/b.3041929/k.BF32/Home.htm.

United Nations Educational, Scientific and Cultural Organization (UNESCO). http://portal.unesco.org.

U.S. Food and Drug Administration, Center for Biologics Evaluation and Research, Department of Health and Human Services. http://www.fda.gov/CBER/ltr/cytotrans070601.htm.

VanderStoep, Scott W., ed. *Science and the Soul: Christian Faith and Psychological Research*. Lanham, MD, and Boulder, CO: University Press of America, 2003.

Wachbroit, Robert. "Normality and the Significance of Difference." In *Wrestling with Behavioral Genetics: Science, Ethics, and Public Conversation*, ed. Erik Parens, Audrey R. Chapman, and Nancy Press, 235–53. Baltimore: Johns Hopkins University Press, 2006.

Waters, Brent. "Disability and the Quest for Perfection: A Moral and Theological Inquiry." In *Theology, Disability, and the New Genetics: Why Science Needs the Church*, ed. John Swinton and Brian Brock, 201–13. London and New York: T&T Clark, 2007.

———. *From Human to Posthuman: Christian Theology and Technology in a Postmodern World*. Burlington, VT: Ashgate, 2006.

Waters, Brent, and Ronald Cole-Turner, eds. *God and the Embryo: Religious Voices on Stem Cells and Cloning*. Washington, DC: Georgetown University Press, 2003.

Watson, James D., and Francis H. C. Crick. "Molecular Structure of Nucleic Acids: A Structure for Deoxyribose Nucleic Acid." *Nature* 171 (April 25, 1953): 737–38.

Weiss, Rick, and Max Aguilera-Hellweg. "The Stem Cell Divide." *National Geographic* 208 (July 2005): 2–27.

West Coast Fertility Centers. http://www.ivfbaby.com.

Wexler, Barbara. *Genetics and Genetic Engineering*. Detroit and New York: Thomson/Gale, 2008.

Wheeler, David A., et al. "The Complete Genome of an Individual by Massively Parallel DNA Sequencing." *Nature* 452, no. 17 (April 2008): 872–76.

Wilmut, Ian, and Roger Highfield. *After Dolly: The Uses and Abuses of Human Cloning*. New York and London: W. W. Norton, 2006.

Wilmut, Ian, A. E. Schnieke, J. McWhir, A. J. Kind, and K. H. Campbell. "Viable Offspring Derived from Fetal and Adult Mammalian Cells." *Nature* 385 (1997): 810–13.

Wilson, Edward O. *Consilience: The Unity of Knowledge*. New York: Alfred A. Knopf, 1998.

———. *The Creation: An Appeal to Save Life on Earth*. New York: W. W. Norton, 2006.

———. *On Human Nature*. Cambridge, MA, and London: Harvard University Press, 2004.

Winterbottom, Michael. *Code 46*. Metro Goldwyn Mayer, 2004.

Wood, Alexis C., et al. "High Heritability for a Composite Index of Children's Activity Level Measures." *Behavior Genetics* 38, no. 3 (May 2008): 266–76.

Yang, Xiangzhong, Sadie L. Smith, X. Cindy Tian, Harris A. Lewin, Jean-Paul Renard, and Teruhiko Wakayama. "Nuclear Reprogramming of Cloned Embryos and its Implications for Therapeutic Cloning." *Nature Genetics* 39, no. 3 (March 7, 2007): 295–302.

Yarri, Donna. *The Ethics of Animal Experimentation: A Critical Analysis and Constructive Christian Proposal*. American Academy of Religion Dissertation Series. Oxford: Oxford University Press, 2005.

Yrigollen, Carolyn M., et al. "Genes Controlling Affiliative Behavior as Candidate Genes for Autism." *Biological Psychiatry* 63, no. 10 (May 2008): 911–16.

Zaninovic, N., J. Hao, J. Pareja, D. James, S. Rafii, and Z. Rosenwaks. "Genetic Modification of Pre-implantation Embryos and Embryonic Stem Cells (ESC) by Recombinant Lentiviral Vectors: Efficient and Stable Methods for Creating Transgenic Embryos and ESC." *Fertility and Sterility* 88, Supplement 1 (September 2007): S310.

INDEX

About the Authors

SPENCER S. STOBER is an associate professor of biology and director of the PhD program in leadership at Alvernia University in Reading, Pennsylvania. Dr. Stober received the Christian R. and Mary F. Lindback Foundation Award for Excellence in Teaching, and he publishes regularly in the *International Journal of Environmental, Cultural, Economic, and Social Sustainability*.

DONNA YARRI is an associate professor of theology and chair of the Department of Humanities at Alvernia University. Dr. Yarri specializes in Christian ethics, regularly teaches courses on medical and global ethics, and has written *The Ethics of Animal Experimentation* (Oxford University Press, 2005).